STEM

Activities
for
Students & Teachers

Membrane Potential,
Simulations,
& Invertebrate Behaviors

Youngwoo Kim
Jiwoo Kim
Elizabeth Womack
Robin L. Cooper

Azalea Art Press
Sonoma · California

ISBN: 978-1-943471-96-6

Science Editor:
Dr. Josh Titlow

Contents

Chapter 1

Resting Membrane Potential & Action Potential with Computational Simulations: High School Through College Level

1.0 Introduction

This chapter covers the following concepts. We describe each concept, offer theoretical exercises, and elaborate on different aspects to consider regarding each concept.

- 1.1. Equilibrium potential
- 1.2. Factors that affect equilibrium potential
- 1.3. Resting membrane potential
- 1.4. Computational simulations with membrane potential
- 1.5. Action potential
- 1.6. Factors that affect membrane potential

Many standard undergraduate textbooks do not address the subtleties needed for a deeper understanding of the factors that impact the resting membrane potential of cells and the shapes of electrical events, such as an action potential in neurons. Although many standard textbooks are designed as an introduction in animal physiology and cover the fundamental basics of the electrical potential of cells and generalized concepts of cellular events that account for electrical events such as an action potential in a neuron, they do not address the finer details. Thus, in this book chapter, the modules are designed to present basic and additional details in neurophysiology concepts for both educators and students to utilize and modify by adding content depending on their needs.

The foundational concepts covered through this chapter can then be used to consider the variations in electrical activity and environmental changes affecting excitable membranes. The basic concept in what produces an electrical potential across a membrane and the various channel types will go beyond the standard models of Na^+ influx (i.e., depolarization) and K^+ efflux (i.e., repolarization, then hyperpolarization) commonly taught for the ionic nature of an action potential. Finally, the parameters presented can be used to learn how to conduct computer simulations of membrane potential based on ionic flow and permeability in various biological conditions.

Below are YouTube links that complement the text to demonstrate the concepts and computational setup.

- Introduction Video:
  https://youtu.be/orwU0zg5Gkk - This video provides an overview of the entire chapter. Students and teachers are encouraged to watch the video at the beginning of the chapter as a preview.
- Full Chapter Video (Detailed):
  https://youtu.be/0NhJ6j9oe1g - This video provides a more detailed explanation of the text.

Below are timestamps that students and/or teachers are encouraged to follow along with.

- o 0:28: Introduction
- o 1:00: Equilibrium & Membrane Potential
- o 1:28: Nernst & Goldman-Hodgkin-Katz Equation
- o 2:22: Action Potential
- o 3:11: Simulations
- o 4:34: Conclusion

1.1. Equilibrium Potential

An electrical potential across a selectively permeable membrane occurs due to chemical difference in concentration and types of ionic charge. This asymmetric distribution of ion concentration and electrical charge is referred to as the **electrochemical gradient**, which gives rise to the potential across membranes of cells. Many model cells have been used to investigate factors which determine the membrane potential of animal cells. Large cells (e.g., large axons of squid nerves, large neurons of a snail, and muscle cells) serve as experimentally favorable models due to the ability to manipulate ionic compositions both within the cell and in the extracellular bathing fluid while monitoring membrane potential via electrodes placed in the cells. Such models allow for investigating how ions move across the membrane at rest and during electrical excitation (Hodgkin and Huxley, 1952a,b; Hodgkin and Katz, 1949; Mullins and Noda, 1963; Thomas, 1969). The skin of an amphibian (i.e. frog) also has historically provided insight into how ions move across membranes of cells (Krogh, 1904, 1937, 1938; Jørgensen et al., 1954; Bruus et al., 1976).

The combination of ion gradients across a biological membrane produces an electrical potential across the membrane. Each ion has an **equilibrium potential** at which movement of the ions reaches a steady state with specific concentrations on each side of the membrane. The derivations of the variables impacting the ionic differences and electrical potential is described by the **Nernst equation** for a single ion.

A generalized cell in a mammal or an invertebrate animal for instructional purposes is shown in **Figure 1**. Generally, a cell has a higher concentration of K^+ ions on the inside and an aggregate of negatively charged compounds as proteins and anions, as compared to the extracellular fluid around the cell. The extracellular fluid then generally has a higher concentration of Na^+, Cl^-, and Ca^{2+} than inside a cell. At a resting state, a cell membrane is generally more permeable to K^+ ions than other ions, which allows K^+ to flow out of the cell when the cell membrane is depolarized from Na^+ or Ca^{2+} ions flowing into the cell.

It is important to note this is not the case for all cells or regions of a cell membrane for a single cell. For instance, some mechanosensory hair cells function with a much higher concentration of K^+ ions outside the cell than inside, thus allowing K^+ to flow into the cell, producing a depolarization at the sensory ending of the cell. Many insect hair cells and hair cells within the mammalian inner ear function in this manner (see Prelic et al., 2022; Fettiplace and Fuchs, 1999). Given the diversity of cell types and

4

morphologies, the generalized cell presented in **Figure 1A** may produce misconceptions. Thus, presenting a cell in figure **Figure 1B** along with **Figure 1A** will keep one's mind open for variations early on when introducing the concepts of membrane excitability. For simplicity, the cell type in Figure 1A is used for the examples in this chapter.

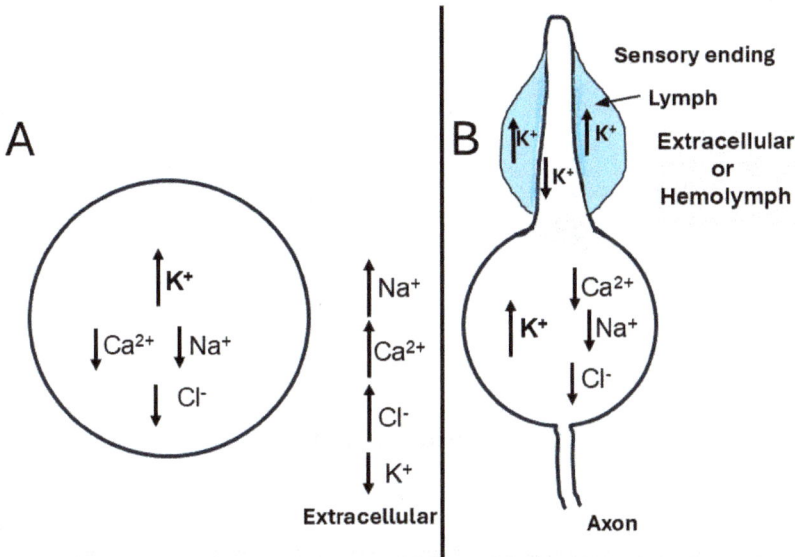

Figure 1: Generalized cells with various concentrations of ions across a membrane. (A) Typical type of cell with ion distribution normally used for teaching purposes. (B) A typical hair cell of an insect with higher K^+ ion concentration in the fluid (lymph or sometimes termed endolymph) surrounding the nerve ending. This fluid is isolated from the hemolymph of an animal's body or the blood or extracellular fluid in a mammal. The fluid around the sensory ending can be different than along the axon of the sensory neuron.

Ions such as K^+ and Ca^{2+}, as well as other anions, establish an electrochemical gradient, which in turn contributes to the equilibrium potential. The **equilibrium potential** can be defined as the electrical difference across a cell due to the concentration of ions. Although equilibrium potentials differ across different cell types, the **Nernst equation** offers a way to estimate equilibrium potential through theoretical calculations. The Nernst equation assumes that ions can move across the membrane (Nernst 1888, 1889) and is defined as follows:

$$V = \frac{RT}{zF} \cdot ln\left(\frac{[X]outside}{[X]inside}\right)$$

Equation 1

- V = equilibrium voltage for the X ion across the membrane
- X = ion of interest
- R = gas constant [8.314 J/(mole•°K)]
- T = absolute temperature [Kelvin]
- Z = valence of the ion
- F = Faraday's constant [9.649×10^4 Coulomb/mole]
- $[X]_o$ = outside concentration of an ion
- $[X]_i$ = inside concentration of an ion

Note that the subscript "o" is for "outside" or extracellular concentration and "i" is for inside or intracellular concentration. These notations will be used through the rest of the text for abbreviation.

Below are examples of calculating the theoretical equilibrium potential using the Nernst equation. Students should be encouraged to plug in the given values into the equation, then check their work following the given values.

Example 1.1.1

Given the following:

- $[K^+]_o = 10$ mM
- $[K^+]_i = 120$ mM
- $T = 21°C = 21° + 273.15°K = 294.15° K$
- 1 Volt = 1 Joule/Coulomb

Place the values into the Nernst equation. Thus, the equation would become the following:

$$E_K = \frac{(\frac{8.314\,J}{mol\cdot °K})(294.15°K)}{(1)(96,490\ Coulomb/mole)} \cdot ln(\frac{10\ mM}{120\ mM})$$

The equation simplifies to the following:

$$E_K = 0.0253 \cdot ln\left(\frac{10\ mM}{120\ mM}\right)$$

$$E_K = 2.303 \cdot 0.0253 \cdot log\left(\frac{10\ mM}{120\ mM}\right)$$

$$E_K = 0.05837 \cdot log\left(\frac{10\ mM}{120\ mM}\right)$$

$$E_K = -62.98\ mV$$

So, -62.98 mV is the equilibrium potential for K^+ given the concentration gradient and temperature.

One can also turn this around and state that if one had a membrane permeable to K^+ and one induced electrical difference with direct electrical potential, as from a battery, across the two sides, a concentration gradient would develop of 10 mM on one side and a 120 mM on the other side.

1.2. Factors That Affect Equilibrium Potential

Equilibrium potential does not always remain constant due to internal and external sources of stimuli. While the Nernst equation provides a single theoretical value, it can still be used to look at how different factors impact equilibrium potential. Different concentrations of ions, as well as temperature, are some well-known factors that can significantly impact equilibrium potential.

Example 1.2.1. Effects of K^+

Knowing the E_k and $[K^+]_O$ one can estimate the $[K^+]_i$ by the following:

- $[K^+]_O$: 10 mM
- Membrane potential: -69.28 mV = 0.06298 V
- T: 294.15° K

Using the Nernst equation (Equation #1),

$$V = \frac{RT}{zF} \cdot ln\left(\frac{[X]outside}{[X]inside}\right)$$

where

$$E_K = \left(\frac{(8.314)(294.15)}{(1)(96490)}\right) ln \left(\frac{10mM}{120mM}\right)$$

$$-0.06298\,V = 2.303 \cdot 0.025345\, log \left(\frac{10mM}{[K^+]_i}\right)$$

$$\frac{-0.06298}{0.05837} = log \left(\frac{10mM}{[K^+]_i}\right)$$

$$antilog(-1.079) = \frac{10mM}{[K^+]_i}$$

$$10^{-1.079} = \frac{10mM}{[K^+]_i}$$

$$[K^+]_i = \frac{10mM}{0.0834}$$

$$[K^+]_i = 119.9 \sim 120mM$$

- R = 8.314 J/(mol•K) (Universal gas constant)
- T = 294.15° K
- z = 1 (Charge of K$^+$)
- F = 96490 C/mol (Faraday's constant)
- 1 Volt = 1 Joule/Coulomb

We substitute the values for back-calculation:

$$E_K = \left(\frac{(8.314)(294.15)}{(1)(96490)}\right) ln \left(\frac{10mM}{120mM}\right)$$

$$-0.06298 \; V \;=\; 2.303 \cdot 0.025345 \; log \left(\frac{10mM}{[K^+]_i}\right)$$

$$\frac{-0.06298}{0.05837} = log \left(\frac{10mM}{[K^+]_i}\right)$$

$$antilog(-1.079) = \frac{10mM}{[K^+]_i}$$

$$10^{-1.079} = \frac{10mM}{[K^+]_i}$$

$$[K^+]_i = \frac{10mM}{0.0834}$$

$$[K^+]_i = 119.9 \sim 120mM$$

This is an estimate of the internal potassium concentration, which is hard to determine experimentally. This is why we went through this exercise. Now, we can do the same process for sodium and other ions to determine their internal concentration values.

Example 1.2.2. Effects of Na⁺

One can derive an equilibrium potential of any ion but recall that this assumes the ion is permeable across the membrane. Try this for Na^+ next:

- $[Na^+]_o = 140$ mM
- $[Na^+]_i = 10$ mM
- T= 21°C

What value did you get?

(A) -66.89 mV
(B) +66.89 mV
(C) +4.77 mV
(D) -4.77 mV

Note that the equilibrium potential value is given in volts, which is 1000x less than the mV value (1 volt = 1000 millivolts).

Example 1.2.3. Effects of Temperature

Now, let's see how temperature affects equilibrium potential. First, let's assume that a normal temperature for a human is 97.5°F = 36.4°C = 309.55°K. By using these values in the equation, we obtain the following values:
- $E_K = -66.28131$
- $E_{Na} = +70.39306$

Both equilibrium potentials increased in value with an increase in temperature, with E_K becoming more negative and E_{Na} becoming more positive.

1.3. Resting Membrane Potential

Equilibrium potential considers the theoretical value of one ion, but membrane potential is the combination of electrical potential established by equilibrium potentials of all ions that can permeate a given cell. **Membrane potential (V_m)** refers to the total electrical difference across a cell due to the influx and efflux of ions (Goldman, 1943; Hodgkin and Katz 1949). Generally, cellular membrane potential does not reside at the same values for any ion's equilibrium potential, which indicates that a mix of ions are responsible for a resting membrane potential and that other factors such as pumps and exchangers contribute to the potential.

For a cell at rest, this difference in electrical charge across the membrane is known as the cell's **resting membrane potential (RMP or RP)**. Considering a living cell, one needs to consider multiple ions and the permeabilities of each ion. This equation is referred to as the **Goldman-Hodgkin-Katz (G-H-K or GHK) equation**. This is also referred to as the constant field equation, considering the field potential across the membrane is constant (Nicholls, et al., 1992, pp.75-76). An important distinction between the GHK and Nernst equation is that the Nernst equation is used only for one specific ion to determine the equilibrium potential for that ion, whereas the GHK equation is used to determine the steady state potential by considering the permeability of multiple ions and their gradients across a cell membrane (Nernst, 1888, 1889; Goldman, 1943; Hodgkin and Huxley, 1952; Hodgkin et al., 1952; Hodgkin and Katz,1949; see Hille, 1992).

A generalized GHK equation for Na^+, K^+, Cl^- and Ca^{2+} ions is as follows:

$$Em=(RT/F)\ Ln\ \frac{\{P_K\,[K]_o + P_{Na}\,[Na]_o + P_{Cl}\,[Cl]_i + P_{Ca}[Ca]_o^{1/2}\}}{\{P_K\,[K]_i + P_{Na}\,[Na]_i + P_{Cl}\,[Cl]_o + P_{Ca}[Ca]_i^{1/2}\}}$$

Equation 2

Since Cl⁻ has a negative charge, the concentration term is inverted in this equation for the inside and outside. This allows the Z (ion charge) to be left off. Given that Ca^{2+} is a 2+ ion, Ca^{2+} concentration is given as a power to $\frac{1}{2}$.

The balance of Na^+ and K^+ across the membrane is maintained by the **Na^+-K^+ ATPase** pump under physiological conditions. Under normal conditions, the pump moves on average, three Na^+ ions out of the cell and two K^+ ions into the cell. This parameter is not considered separately in the GHK for maintaining the potential. This Na^+-K^+ ATPase pump is also considered **electrogenic**, as it has a greater ability to pump when the membrane is depolarized (Skou, 1989a,b; Thomas, 1969). In many cells, the pump speeds up when a cell is electrically activated by depolarization.

(Fun fact: A Nobel Prize in chemistry was awarded in 1997 for this discovery made back in the late 1950s. The fundamentals of the discovery were obtained from research using axons from a crab (Skou, 1965, 1998)).

In addition, there are various pumps (i.e., plasma membrane Ca^{2+} pump referred to as the PMCA) and various ion exchangers on the membranes of cells (i.e., Na^+-Ca^{2+} exchanger referred to as NCX) that are also not considered in the GHK equation (Moody, 1981; Thomas, 1977; see Nicholls, Martin and Wallace, 1992, pp. 66-89 for review). **Ion channels** (both leak and voltage gated) and **pumps/exchangers** can vary in function due to environmental conditions such as pH, temperature, and ionic environment (Cheslock et al., 2021, Ruff, 1999;

Feliciangeli et al., 2015; Mert et al., 2003), which would influence membrane potential.

Another type of equation that is widely used to measure the membrane potential is the **flux equation**. The GHK flux equation is a modified version of the GHK voltage equation. The voltage equation assumes a steady state of the ion concentration values, meaning that it describes the flux when the membrane potential is at a constant value. The flux equation, on the other hand, is during dynamic changes like an action potential; the flux equation describes the movement of ions across a cell membrane by considering both the concentration gradient and the transmembrane potential. However, one can use the GHK equation to estimate values along with dynamic changes in increments. Various modifications have been implemented over the years to the GHK equation, one being the **Poisson-Nernst-Planck ion channel model**. This approach considers the charges of the ion channel and ionic interactions in the medium (Gardner et al., 2004; Liu, 2009; Xie and Lu, 2020). However, the GHK and Poisson-Nernst-Planck equations both fail to consider factors like passive ion exchangers and the electrogenic Na^+-K^+ pump (Hernandez et al. 1989). Given all the effort over the years it has not been possible with mathematical estimations to describe the membrane potential fully in varied conditions. Another concept is **Ling's adsorption theory**. It proposes that membrane potential and action potentials in living cells arise from the adsorption of mobile ions onto specific sites on the cell membrane, rather than from transmembrane ion transport (Ling, 1992; Tamagawa and Morita, 2014). Ling's idea is commonly cited in literature, but it is not widely used. These other approaches are mentioned here so the reader is aware that they exist.

One can calculate the potential by considering various ions and their permeabilities using the GHK equation.

Example 1.3.1.

Consider a cell with RP around -50 mV and two ions, K^+ and Na^+. Assuming E_K = -80 mV, E_{Na} = +35 mV, the resting membrane potential will be close to E_K because the cell is more permeable to K^+ (P_K) than to Na^+ (P_{Na}).

The passive protein channels that mediate ion flux are called **leak channels**. The Na^+ leak channel is known as **NALCN** and the K^+ leak channel is termed as a **K2P** (i.e., two-pore-domain K^+ channel) (Goldstein et al., 1998; Plant and Goldstein, 2015; Monteil et al. 2024). There are many types of K2P channels; in fact, nearly all cells express a variety of K2P subtypes. They have different properties and sensitivities to pH, temperature, and stretch of the membrane; so, all these factors could impact ion flux across the membrane. However, let's keep it simple first and just think about the driving gradients for K^+ and Na^+: consider why the resting membrane potential is not at E_K and how RP varies with $[K^+]_O$ being raised. We can also for now group all the subtypes of K2P channels to be represented with a single value for P_K. We will then use the GHK equation for Na^+ and K^+.

Example 1.3.2: Crayfish Muscle

Below are values for GHK, as determined from the literature, except for $[Ca^{2+}]_i$ and P_{Ca}, which were added to represent the full GHK as shown above.

For crayfish muscle: "*Procambarus* and *Astacus* $[K^+]_i$ appears to be 171 and 167 mM, respectively." (Katz et al., 1972; Atwood, 1982). Here the values are for *Procambarus clarkii*:

$[K^+]_i$ = 171 mM (determined for crayfish muscle)
$[K^+]_O$ = 5.3 mM (Saline)

$[Na^+]_i$ = 17.4 mM (assume for muscle)
$[Na^+]_O$ = 205 mM (Saline)

$[Cl^-]_i$ = 12.7 mM (assume for muscle)
$[Cl^-]_O$ = 232.15 mM (assume from saline made with: 205 mM NaCl; 5.3 mM KCl; 13.5 mM $CaCl_2 2H_2O$; 2.45 mM $MgCl_2 6H_2O$)

$[Ca^{2+}]_i$ = 0.00001 mM (assume for free ion in muscle)
$[Ca^{2+}]_O$ = 13.5 mM (Saline)

P_K = 1 (assume for muscle)
P_{Na} = 0.001 (assume for muscle)
P_{Cl} = 0.01 (assume for muscle)
P_{Ca} = 0.00001 (assume for muscle)

Note that it is assumed that there is an aggregate of the K2P subtypes and NALCN channels that are open accounting for P_K and P_{Na} but not accounting for pumps and exchangers.

Let's work out a simulation with the GHK equation and put in some values for ions.

1.4. Computational Simulations with Membrane Potential

To program the simulation, one will need to set up a Python integrated environment. There are three main components to achieving this:

1. Python Interpreter
2. Integrated Development Environment (IDE)
3. Python Extension

Step 1. Install a Python Interpreter on your computer through this site: https://www.python.org/downloads/.

- If one is a Windows user, all one needs to do is click the "Download for Windows" button for the latest version.

- If one is a MacOS user, Python is not supported. A package management system is required (e.g. Homebrew). To install Python using Homebrew, type "brew install python3" on the Terminal prompt. To check if Python is installed onto the Mac, type the command "python3 --version" on Terminal. If Python is installed, one will receive a message that states which version one has. If Python is not installed, one will see "command not found: python" on the terminal.

```
● ● ●                       🖥 ywkim17 — -zsh — 80×24
Last login: Tue Mar 18 12:50:13 on ttys001
ywkim17@gim-yeong-us-Laptop ~ % brew install python3
```

```
● ● ●                       🖥 ywkim17 — -zsh — 80×24
Removing: /Users/ywkim17/Library/Caches/Homebrew/pkg-config--0.29.2_3... (235.9K
B)
Removing: /Users/ywkim17/Library/Caches/Homebrew/portable-ruby-3.3.5.arm64_big_s
ur.bottle.tar.gz... (11.2MB)
Removing: /Users/ywkim17/Library/Caches/Homebrew/portable-ruby-3.3.4_1.arm64_big
_sur.bottle.tar.gz... (11.1MB)
Removing: /Users/ywkim17/Library/Caches/Homebrew/pkg-config_bottle_manifest--0.2
9.2_3... (13.1KB)
==> Caveats
==> python@3.13
Python is installed as
  /opt/homebrew/bin/python3

Unversioned symlinks `python`, `python-config`, `pip` etc. pointing to
`python3`, `python3-config`, `pip3` etc., respectively, are installed into
  /opt/homebrew/opt/python@3.13/libexec/bin

See: https://docs.brew.sh/Homebrew-and-Python
==> datalad
zsh completions have been installed to:
  /opt/homebrew/share/zsh/site-functions
ywkim17@gim-yeong-us-Laptop ~ % python3 --version
Python 3.12.6
ywkim17@gim-yeong-us-Laptop ~ %
```

Step 2. Download an Integrated Development Environment (IDE) for the code to be compiled on. Visual Studio Code (VSCode) is a commonly used IDE. One can download it here:

https://code.visualstudio.com/download

Step 3. Install a Python Extension onto the IDE by clicking on the extensions icon on the left sidebar of the VSCode platform. Search "Python" and download the extension.

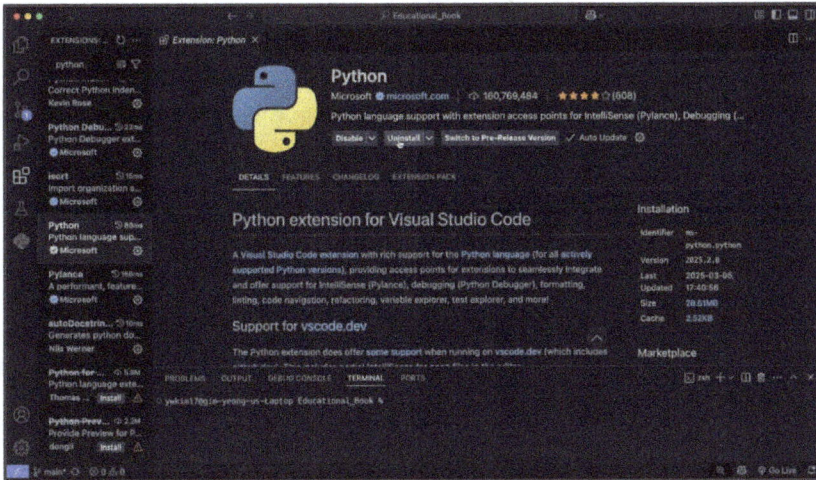

Now, one can go to the explorer and create a new Python file with a ".py" ending. One is now ready to program!

Running the Nernst Equation File

If one has not already, create a GitHub account (there is no cost). Once logged in, go to this link: https://github.com/ywkim17/Educational_Book/blob/main/Nernst.py.

Download the code named "Nernst.py" and run it on the VSCode software. One can now run simulations through one's computer!

The Nernst.py code is shown below:

```
import math
def nernst_equation(R, T, F, z, ion_in, ion_out):
    """

    Computes the Nernst equilibrium potential (E_ion) in millivolts.
    Parameters:
        R     : float - Universal gas constant (J/(mol·K))
        T     : float - Temperature in Kelvin (K)
        F     : float - Faraday's constant (C/mol)
        z     : float - Charge of the ion (e.g., +1 for K⁺, Na⁺; -1 for Cl⁻)
        ion_in : float - Intracellular ion concentration (mM)
        ion_out : float - Extracellular ion concentration (mM)

    Returns:
        E_ion (float): Equilibrium potential in millivolts (mV)
    """
    if ion_in <= 0 or ion_out <= 0:
        raise ValueError("Ion concentrations must be greater than zero.")
    # Calculate RT/zF in millivolts
    V_T = (R * T) / (z * F) * 1000  # Convert to mV
    E_ion = V_T * math.log(ion_out / ion_in)
    return E_ion
# User Inputs
R = 8.314
T = 310
F = 6485
# Example Ion Data
z = 1
ion_in = 200
ion_out = 5
# Compute Nernst Potential
E_ion = nernst_equation(R, T, F, z, ion_in, ion_out)
print(f"\nEquilibrium potential: {E_ion:.2f} mV")
```

Running the GHK Simulation File

If one has not already, create a GitHub account. Once logged in, go to this link:

https://github.com/ywkim17/Educational_Book/blob/main/GHK.py.

Download the code named "GHK.py" and run it on the VSCode software. One can now run simulations through one's computer!

The GHK.py code is shown below:

```python
import math
def ghk(R, T, F, P_K, K_in, K_out, P_Na, Na_in, Na_out, P_Cl, Cl_in, Cl_out, P_Ca, Ca_in, Ca_out):
    """
    Computes membrane potential (Vm) in millivolts using the Goldman-Hodgkin-Katz equation.
    Parameters:
        R     : float - Universal gas constant (J/(mol·K))
        T     : float - Temperature in Kelvin (K), room temperature = 273.15 + 21 = 294.15
        F     : float - Faraday's constant (C/mol)
        P_K   : float - Permeability of potassium
        K_in  : float - Intracellular potassium concentration (mM)
        K_out : float - Extracellular potassium concentration (mM)
        P_Na  : float - Permeability of sodium
        Na_in : float - Intracellular sodium concentration (mM)
        Na_out: float - Extracellular sodium concentration (mM)
        P_Cl  : float - Permeability of chloride
        Cl_in : float - Intracellular chloride concentration (mM)
        Cl_out: float - Extracellular chloride concentration (mM)
        P_Ca  : float - Permeability of calcium
        Ca_in : float - Intracellular calcium concentration (mM)
        Ca_out: float - Extracellular calcium concentration (mM)

    Returns:
        Vm (float): Membrane potential in millivolts (mV)
    """

    # Calculate RT/F in millivolts
    V_T = (R * T) / F * 1000  # Convert to mV

    numerator = (P_K * K_out + P_Na * Na_out + P_Cl * Cl_in + P_Ca ** Ca_out[1] )
    denominator = (P_K * K_in + P_Na * Na_in + P_Cl * Cl_out + P_Ca ** Ca_in)

    if denominator == 0:
        raise ValueError("Denominator is zero; check input values.")

    Vm = V_T * math.log(numerator / denominator)
    return Vm
# Parameter Inputs
R = 8.314 #Constant Value
T = 294.15 #Room Temperature in Kelvin
F = 96485 #Constant Value
P_K, K_in, K_out = 1.0, 140, 5
P_Na, Na_in, Na_out = 0.04, 15, 145
P_Cl, Cl_in, Cl_out = 0.45, 10, 120
P_Ca, Ca_in, Ca_out = 0.01, 0.0001, 1.8
# Compute Membrane Potential
Vm = ghk(R, T, F, P_K, K_in, K_out, P_Na, Na_in, Na_out, P_Cl, Cl_in, Cl_out, P_Ca, Ca_in[2] , Ca_out)
print(f"\nMembrane potential: {Vm:.2f} mV")
```

Example 1.4.1. K2P Channel Overexpression

Let's change the values to assume that one had an overexpression of, for example, the K2P channels. Then, the probability would be increased; let's try that out with some numbers. We want to keep with those numbers and change one parameter (temperature). Predict what would happen to the resting membrane potential with an increase in temperature. Increase it and see how the value changes. Then, decrease it and see how the value changes

What happens to the resting membrane potential when temperature increases?

(A) It depolarizes.
(B) It hyperpolarizes.
(C) It remains unchanged.

> **Answer: (C) It hyperpolarizes.**

Try one of the GHK simulations and only vary temperature and note how the V_m is altered.

Now, one understands how temperature might affect the resting membrane potential. It's mostly related to the K^+ permeability because that's the highest permeability, but it also affects the equilibrium for sodium. If one had a higher conduction of Na^+, then that would also drive membrane potential away from the negative value of the equilibrium potential for K^+.

When one considers opening a voltage-gated ion channel, the driving gradient for that ion depends on the equilibrium potential for that ion. So as shown, temperature changes equilibrium potential and, thus, membrane potential when voltage-gated channels open.

Example 1.4.2. Effect of $[Ca^{2+}]_O$

Another very interesting phenomenon is the effect of $[Ca^{2+}]_O$ on membrane potential. The interesting aspect is that not all preparations behave the same way to the effect of $[Ca^{2+}]_O$. Let's consider the values for the crayfish muscle used above but change $[Ca^{2+}]_O$ = 13.5 mM (Saline) to 23 mM. Physiological experiments to measure the membrane potential of crayfish muscles are relatively easy to conduct due to the large cells and minimal ingredients of the physiological saline along with being able to conduct recordings at room temperature (Baierlein et al., 2011).

How does the V_m change, given these conditions below?

$[K^+]_i$ = 171 mM (determined for crayfish muscle; Atwood, 1982)
$[K^+]_O$ = 5.3 mM (Saline)

$[Na^+]_i$ = 17.4 mM (assume for muscle)
$[Na^+]_O$ = 205 mM (Saline)

$[Cl^-]_i$ = 12.7 mM (assume for muscle)
$[Cl^-]_O$ = 232.15 mM (assume from saline made with: 205 mM NaCl; 5.3 mM KCl; 13.5 mM CaCl$_2$ 2H2O; 2.45 mM MgCl$_2$ 6H2O)

$[Ca^{2+}]_i$ = 0.00001 mM (assume for free ion in muscle)
$[Ca^{2+}]_O$ = 13.5 mM (Saline)

P_K = 1 (assume for muscle)
P_{Na} = 0.001 (assume for muscle)
P_{Cl} = 0.01 (assume for muscle)
P_{Ca} = 0.00001 (assume for muscle)

As expected here, there was not much change, and that is what is found experimentally for crayfish muscle as well (de Castro et al., 2025, preliminary data). However, for larval *Drosophila* muscle, sensory neurons in crayfish and crab, and motor neurons in crayfish, this is not necessarily the case. It has been shown that the resting membrane potential in larval *Drosophila* muscle is influenced by [Ca^{2+}] (Krans et al., 2010; Elliott and Cooper, 2024).

Let's now consider empirical data obtained by electrophysiological measures and make some predictions from parameters through simulations to explain the observations shown in **Figure 2**.

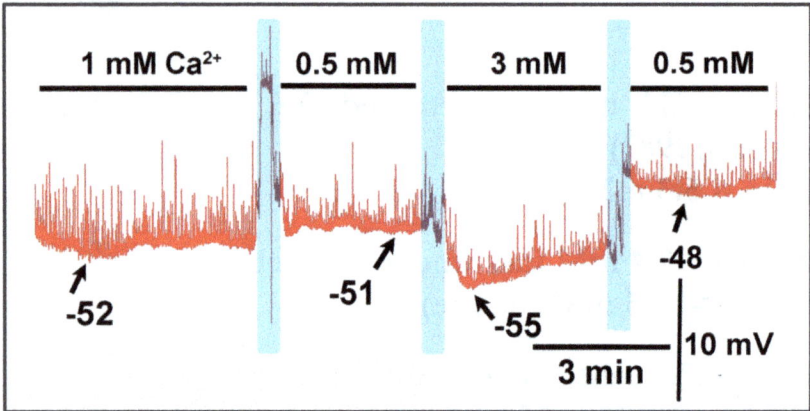

Figure 2: *The responses obtained from larval Drosophila muscle 6 while altering extracellular [Ca^{2+}]. The upward deflections are the responses from vesicles fusing within the presynaptic nerve terminal resulting in quantal postsynaptic events. The blue boxes indicate when the bathing solution was exchanged. Obtained from Elliott and Cooper (2024).*

Example 1.4.3. Larval Drosophila Muscle

For the larval *Drosophila* muscle, let's use **Equation #2** and the following parameters:

The saline used for larval *Drosophila* physiological measures is (in mmol/L) 70 NaCl, 5 KCl, 20 $MgCl_2$ 6 H20 10 $NaHCO_3$, 1 $CaCl_2$, 5 trehalose, 115 sucrose, 25 N,N-bis(2-hydroxyethyl)-2-aminoethane sulfonic acid (BES) and pH at 7.2.

$[K^+]_i$ = 190 mM (estimated for Drosophila muscle)
$[K^+]_O$ = 5 mM (Saline)

$[Na^+]_i$ = 12 mM (assume for Drosophila muscle)
$[Na^+]_O$ = 80 mM (Saline)

$[Cl^-]_i$ = 12 mM (assume for muscle)
$[Cl^-]_O$ = 91 mM

$[Ca^{2+}]_i$ = 0.00001 mM (assume for free ion in muscle)
$[Ca^{2+}]_O$ = 1 mM (Saline)

P_K = 1 (assume for muscle)
P_{Na} = 0.001 (assume for muscle)
P_{Cl} = 0.01 (assume for muscle)
P_{Ca} = 0.00001 (assume for muscle)

Now run the simulation with 1 mM $[Ca^{2+}]_O$ and again with 3 mM $[Ca^{2+}]_O$. Did the V_M change? Now let's try changing one other parameter by using $[Ca^{2+}]_O$ and P_{Na} from 0.001 to 0.0005. What was the result on the V_m?

-91.33 to -91.52.

Unusual as it might seem, $[Ca^{2+}]_O$ blocks NALCN channels, which account for altering the P_{Na} at rest (Krans et al., 2010; Monteil et al., 2024; Armstrong and Cota, 1999; Lu et al., 2007) and likely throughout any electrical events such as an action potential.

Computationally, let's work through 4 conditions:

Normal resting membrane with 1 mM $[Ca^{2+}]_O$ and P_{Na} at 0.001
V_m = -91.33

Then reduced $[Ca^{2+}]_O$ to 0.5 and P_{Na} at 0.0015

$$V_m = -91.14$$

Then reduced $[Ca^{2+}]_O$ to 0.1 and P_{Na} at 0.002

$$V_m = -90.94$$

Lastly, reduced $[Ca^{2+}]_O$ to 0 and P_{Na} at 0.005

$$V_m = -89.82$$

Notice in **Figure 3** that when $[Ca^{2+}]_O$ is reduced, so is the E_{Ca}. Also, notice the resting membrane potential theoretically depolarizes in unison. So, depending on the distribution and density of the NALCN channels on a membrane, the effects of $[Ca^{2+}]_O$ will vary. We have observed that the $[Ca^{2+}]_O$ has little effect on the resting membrane potential on the walking leg opener muscle in crayfish but has a significant effect on the larval *Drosophila* muscle (m6). We would expect possible differences in the type of NALCN channel (i.e. an isoform difference) between the two species or a difference in the protein complex associated with the NALCN channel (Lee et al. 1999; Cochet-Bissuel et al., 2014; review in Monteil et al. 2024).

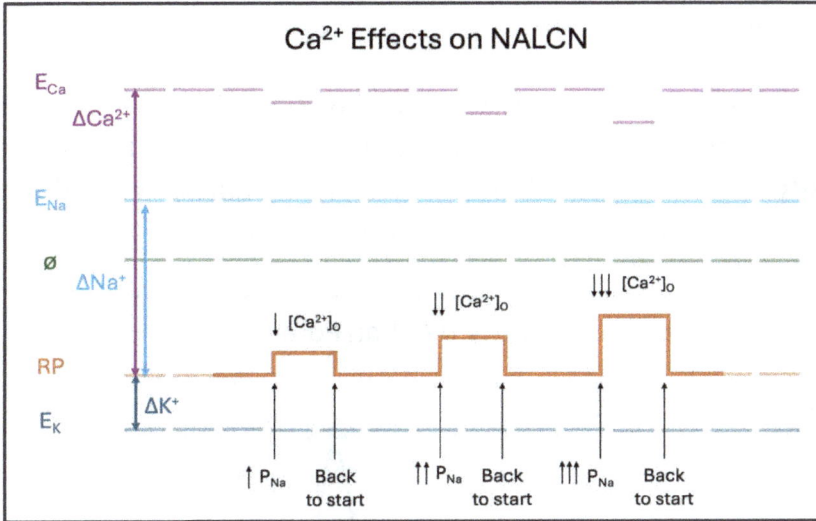

Figure 3: *Schematic representation of the equilibrium potentials and the effect of reducing [Ca^{2+}]$_O$ on the resting membrane potential (RP). The theoretical effects in reducing the block by Ca^{2+} on the NALCN channel for this model allows the membrane potential to depolarize due to the increased Na$^+$ leak through NALCN channels. The arrows indicate the various driving gradients of the particular ions at the resting membrane potential. They would change accordingly with the changes in the RP.*

Now we want to start with a standard shaped action potential in a neuron and work through how that shape is produced by the various fluxes of ions.

We will take this in steps only using Na$^+$ and K$^+$ ions to start with. This is adding all the types of voltage gated Na$^+$ and K$^+$ channels which could be opening, as well as including the leak channels for K$^+$ and Na$^+$ ions, as these leak channels are constitutively open.

1.5. Action Potential

The membrane potential may undergo changes in values, such as through the action potential. The **action potential** is a rapid change in the membrane potential that allows for the transmission of signals between cells.

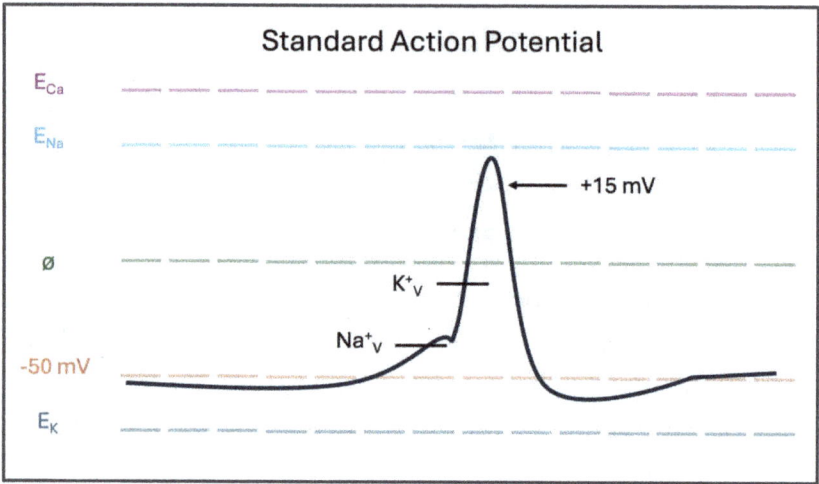

Figure 4: *A generalized shape of an action potential illustrating the driving gradients to the various equilibrium potentials for E_K, E_{Na}, E_{Ca}. When voltage gated channels sense their threshold the majority of the channels of that subtype will open. $Na^+{}_V$ and $K^+{}_V$ refers to voltage gated ion channels. The gradual rise indicates a few $Na^+{}_V$ channels opening and then at the line where the voltage threshold occurs for the majority of the channels. The threshold for the $K^+{}_V$ occurs at a more depolarized level in this example.*

Now try to recreate a standard shaped action potential in key points along the outline of an action potential to gain sense of how much the permeability is changing for the ions at various points during the duration of the action potential. We will try to recreate the upstroke of the action potential, which goes from RP (~ -50mV) to a peak of the action potential (~ +15 mV). What is generally shown in textbooks is gradual depolarization and then threshold is obtained for a majority of the voltage Na^+ channels ($Na^+{}_V$), which produces the rapid upstroke of the action potential. In **Table 1,** there are some values and the representative points for a theoretical action potential (**Figure 5**).

Constant Values:
- $[K^+]_i = 140$
- $[K^+]_o = 5$
- $[Na^+]_i = 15$
- $[Na^+]_o = 145$
- Room temperature (294.15° K)

Points	P_K	P_{Na}	mV measured
1	1	0.1	-53.15
2	1	0.2	-41.37
3	1	0.3	-33.46
4	1	0.4	-27.52
5	1	0.6	-18.84
6	0.8	0.8	- 9.06
7	0.5	1	2.27
8	0.1	1	15.00
9	0.1	0.8	10.49
10	0.1	0.3	-10.18
11	0.5	0.3	-23.66
12	1	0.2	-41.37
13	2	0.1	-62.04
14	1	0.1	-53.15

Table 1: The values to place into the GHK equation (Eq.#2) to obtain various points as indicated in the schematic below for a representative action potential. Notice that the P_{Na} drastically increases after the threshold is reached until the $Na^+{}_V$ channels start to show inactivation. Likewise, P_K increases after the threshold is reached for $K^+{}_V$ channels and remains high until the membrane potential regains the original resting membrane potential.

After obtaining the mV values for each step, one now needs to graph the shape of the action potential in a graphing software like Excel. The X-axis plot can vary to make the action potential wider or narrower. We suggest a plot like this below with the associated Excel data table.

	A	B	C
1	Points on graph	Arranged time	mV
2	From simulation	Sample below	Values
3		1	0 ?
4		2	
5		3	
6		4	
7	...		

Table 2: The values used to graph the figure shown in Figure 4. Notice we spread out the values in the X axis to better view the shape of the action potential.

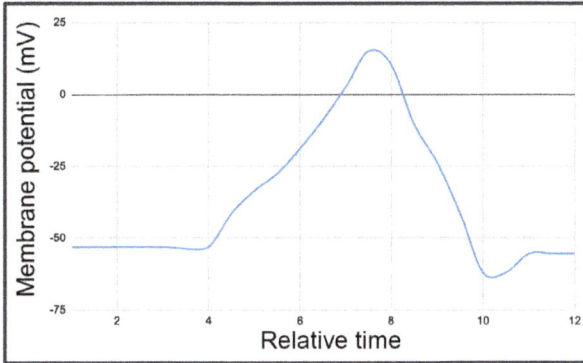

Figure 5: *The representative points plotted for the theoretical action potential from values listed in Table 1. The numbers shown correspond to the column numbers in Table 1.*

In the classic experiments by Hodgkin and Katz (1949) using the squid axon, they determined that the upstroke and amplitude of the action potential was due to Na^+ ion influx. To demonstrate this concept, they reduced extracellular concentration of Na^+ (i.e. $[Na^+]_O$). The same general effect is also observed for motor nerve axons in crayfish (Desai-Shah and Cooper, 2010).

Example 1.5.1. Varying $[Na^+]_O$

The effect of reducing the $[Na^+]_O$ in steps is illustrated in **Figure 6**. As an exercise use the values in **Table 1** but reduce the $[Na^+]_O$ from 145 mM to two different lower values.

List peak amplitudes obtained for point 8 in Table 1 for the following and fill in the blanks.

Start $[Na^+]_O$ = 145 mM; the peak amplitude is 15.45 (-67.90 mV —- at point 8 this is + 15 mV)

1st reduction in $[Na^+]_O$ = 100 mM; the peak amplitude is 6.72 ($[Na^+]_O$ = 100mM, -71.24 mV)

31

2nd reduction in $[Na]_O = 50$ mM; the peak amplitude is - 8.74 ($[Na^+]o = 50mM$, -75.52 mV)

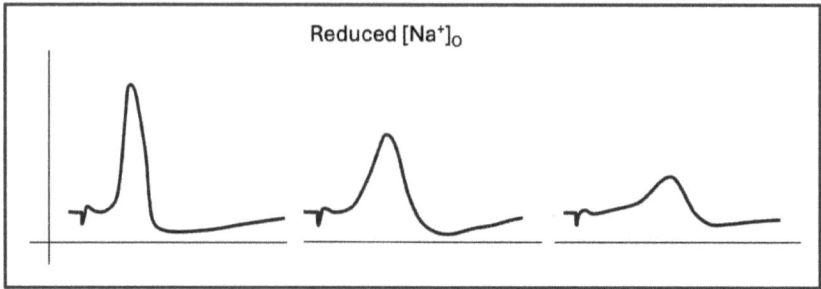

Figure 6: *Theoretical shapes of action potentials as the $[Na^+]o$ is reduced from representative traces obtained using a squid axon. Modified from Hodgkin and Katz, (1949, Figure 3).*

As it is now becoming clear that the external concentrations of ions can have a large impact on the shape of the action potential, so does the distribution and density of ion channels. The presynaptic endings of nerve terminals are the locations where synaptic vesicles fuse to release their contents for chemical synaptic transmission. This is generally associated with close association with **voltage-gated Ca^{2+} channels** (Ca^{2+}_V) to the docking sites of synaptic vesicles (Matthews, 1996). These Ca^{2+}_V open when the terminal is depolarized by the arrival of an action potential. The exercise now is to determine how the shape of the action potential would theoretically change in shape, particularly the amplitude of the action potential. The width would likely also change due to a larger amplitude event. But for this exercise, one can compare the amplitude without and with the Ca^{2+}_V opening.

Use Table 1 but now modify for the contribution of Ca^{2+}_V opening using:

$[Ca^{2+}]_i = 0.00001$ mM
$[Ca^{2+}]_O = 13.5$ mM (Saline)

Points	pK	pNa	pCa	pCa Terminal	mV measured	mV measure with Ca^{2+}_V opening
1	1	0.001	0.0001	0.0001	-80.44	—
2	1	0.0015	0.0001	0.0001	-80.37	—
3	1	0.002	0.0001	0.0001	-80.30	
4	1	0.1	0.0001	0.0001	-69.49	
5	1	1.0	0.0001	0.01	-33.56	
6	0.5	1.0	0.0001	0.01	-23.81	
7	0.1	1.0	0.0001	0.001	-10.99	
8	0.1	1.5	0.0001	0.0015	-3.29	
10	0.01	2	0.0001	0.0001	5.40	

Table 2: These are parameters to use to start off the simulations of the action potential shape in an axon and in the axon terminal with an increase in P_{Ca}.

Take the mV values as the Y-axis without the P_{Ca} changing and graph in Excel of some graphing program, then graph on top of this graph with different symbols or colors the mV values obtained with P_{Ca} changing.

A representative action potential in a presynaptic nerve terminal may appear with a larger amplitude and a wider shape. Compare **Figure 4** above in an axon with **Figure 7** at the nerve terminal. As an exercise in investigating the scientific literature, try to find any primary research article where the shape of the action potential has been measured in the presynaptic nerve terminal of any mammalian neuromuscular junction. This will enhance one's appreciation of using invertebrate models for making such direct measures. By trying to find this information in the literature, one will then gain an appreciation that by

33

inference, one can apply such knowledge obtained in invertebrate preparations to predict what is likely occurring in mammalian preparations. Such measures are able to be made in the pre-terminal region of crayfish motor neurons.

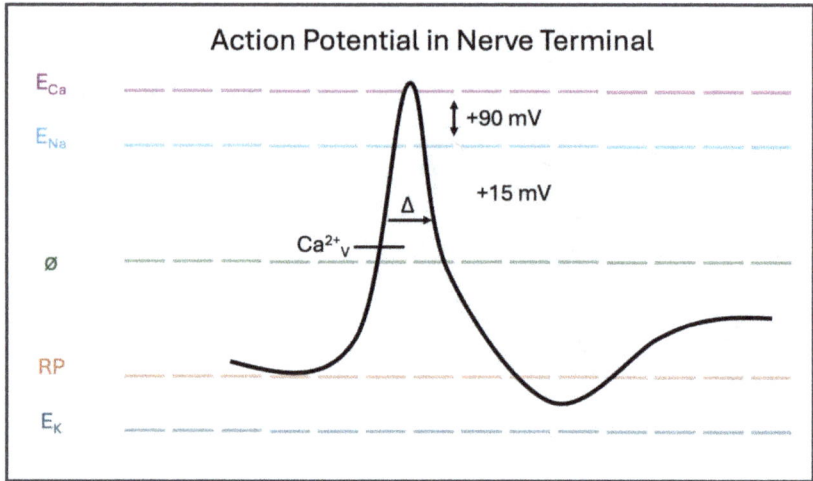

Figure 7: *A hypothetical shape of an action potential in a presynaptic terminal of a neuron, where Ca^{2+}_V channels are in a high density for aiding in chemical transmission.*

When considering the conductance of ions and their driving gradients during an action potential, it is easy to forget that the **leak channels** are also contributing to the shape of the action potential, as they are constantly open. The further away from the equilibrium potentials for those ions, the greater the driving force for the conductance. Likewise, the closer the membrane potential is to the equilibrium potential the driving force is reduced. To illustrate this, a schematic hypothetical illustration is provided. The conductance for the aggregate of K2P subtypes and NALCN leak channels along with two generalized and aggregated K^+_V and Na^+_V voltage-gated conductance's are shown as relative areas under the curve relating to the amount of conductance during an action potential (**Figure 8**).

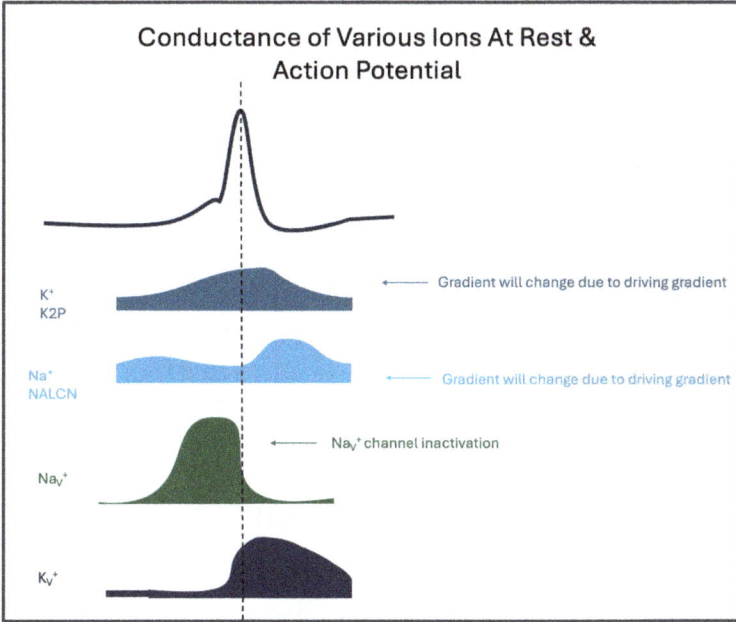

Figure 8: *Representative conductance's of the ions for the leak channels (K2P and NALCN), along with the generalized Na^+ and K^+ conductance for the voltage gated ion channels at rest and during an action potential. The relative area under the curve corresponds to the degree of conductance for the various aggregated subtypes of channels.*

In considering the generalized action potential as represented in **Figure 4** one subtype of each K^+_V and Na^+_V channel is assumed. However, in many cells, the action potential is really composed of a mix of K^+_V and Na^+_V subtypes, which have different thresholds of activation and various kinetics to the channels affecting the overall conductance of the ions and the shape of the action potential (Tsantoulas and McMahon, 2014). A schematic representation of 3 types of K^+_V channels with varied thresholds of activation are illustrated in **Figure 9**. To illustrate the effect on the shape of the action potential by different types of K^+_V, an exercise in superimposing various plots from the simulations will emphasize this point. Also, it will be entertaining to add or delete various types of channels to see the effects. This requires a modification of the GHK equation to account for the various subtypes. This slight modification is not normally addressed in textbooks, but we feel it allows one to quickly address the content in how subtypes of ion channels can influence the shape of the action potential. A table is provided with 3 different subtypes of K^+_V and one Na^+_V.

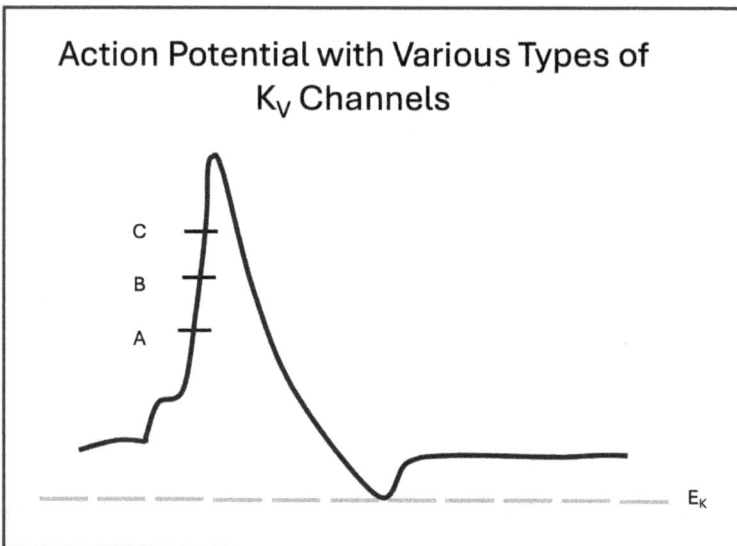

Figure 9: *A representative action potential with three different subtypes (A, B and C) of K^+_V channels with varied threshold of activation and kinetics (modified from Tsantoulas and McMahon, 2014).*

Three subtypes of K^+_V are represented in **Figure 9** as A, B and C are shown for the type of K$^+$ current that would occur for a depolarizing square wave to activate the channels. One subtype that opens quickly is termed **fast-activating**, which does not show inactivation (**Figure 10 A**). Another that is fast-activating but shows inactivation (**Figure 10B**), and a third type that is slow-activating (**Figure 10C**). **Figure 11** is an overview of these three subtypes during an action potential which will be simulated in this next exercise.

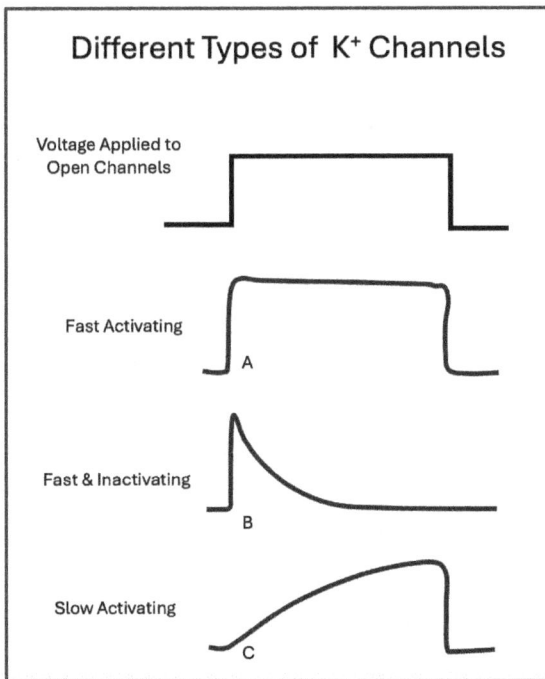

Figure 10: *Three subtypes of K^+_V channels which have varied thresholds of activation and kinetics upon being stimulated to open. (A) One subtype that is fast activating and does not show inactivation. (B) A fast-activating subtype but shows inactivation. (C) A slow activating subtype. (Modified from data shown in Tsantoulas and McMahon, 2014).*

A generalized GHK equation for Na^+, K^+, modified to account for the 3 types of K^+_V channels as follows:

$$Em=(RT/F)\ Ln\ \frac{\{(P_K A + P_K B + P_K C)\ [K]_o + P_{Na}\ [Na]_o\}}{\{(P_K A + P_K B + P_K C)\ [K]_i + P_{Na}\ [Na]_i\}}$$

Equation 3

Use a table format as above but now use P_K A, P_K B, and P_K C at various time points during the theoretical action potential. These values do not consider Ca^{2+} ions.

Points	pKA	pKB	pKC	pNa	mV with ABC	mV with AC	mV with BC
1	.8	.02	0	0.001	-75.03	-74.90	-62.93
2	1.8	.05	0	0.0015	-78.82	-78.70	-63.58
3	2	.1	0.01	0.002	-79.23	-79.03	-65.01
4	2	.5	0.05	0.1	-65.00	-62.36	-45.80
5	2	.9	0.1	0.2	-57.94	-52.63	-41.38

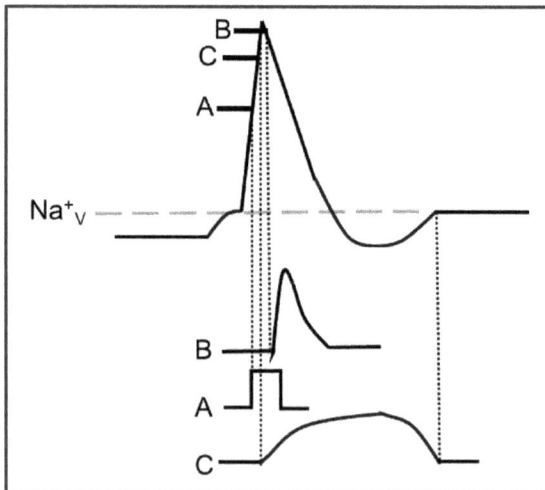

Figure 11: *The threshold of opening and contributions of the three K^+_V channels shown in Figure 10 to the shape of a theoretical action potential.*

Now, we want to plot the mV obtained for the three conditions in different symbols or colors and connect the points. The timeline on the Excel table provided below would help to spread out the action potential for better viewing.

	A	B	C	D
1	Time	mV with ABC	mV with AC	mV with BC
2	0			
3	????			
4				
5				
6				
7				
8				
9				
10				
11				
12				
13				
14				
15				
16				
17				
18				
19				

Table 4: *A representative data table from values listed in Table 3 now to be used with the provided time points for the X-axis to depict the effects of varied subtypes of K^+_V channels on the shape of an action potential.*

Plot the three graphs, and discuss the varied shapes in the action potential that occur by adding or subtracting various subtypes of K^+_V channels. A potential graph as shown on the next page might be representative (**Figure 12**).

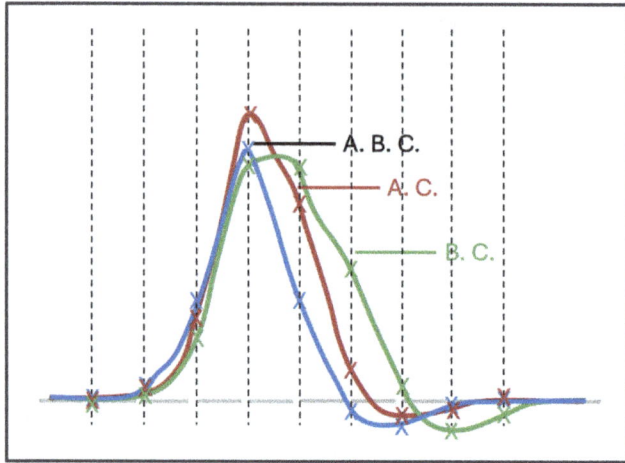

Figure 12: *A potential figure that could be generated from graphing the values in Table 4 with varied type of voltage gated* K^+ *channels.*

The shape of action potentials is generally assumed to be uniform in various regions of a neuron in examples used in educational material. However, now working through the various exercises and simulations, one is realizing the immense variation in the shape of an action potential even within a single neuron depending on the location within a neuron. To illustrate this variation, a generalized figure (**Figure 13**) modified from Tsantoulas and McMahon (2014) shows variation in the distribution and density of the subtypes in K^+_V, Na^+_V, K2P channels, along with NALCN channels located along unmyelinated and myelinated representative neurons. This figure represents sensory neurons relaying information from the skin to the spinal cord of a vertebrate animal. It is feasible to assume the various shapes of action potentials in the various regions but difficult to know really without information on all the variables (i.e., density and types of channels, as well as membrane resistance). Only with direct intracellular measures across the membrane would one know for sure.

A wonderful example of this type of experiment was performed with the motor neuron innervating the walking leg opener muscle in the crayfish (Lin, 2016). The amplitude of the action potential decreased along the axon to the presynaptic terminal and was still able to conduct an action potential in a Na^+-free environment. This implied the action potential was initiated more proximal along the axon and conducted via cable properties to the nerve terminal. There is some Na^+ influx along the distal aspects of the axon but not enough to rejuvenate or produce new action potentials, but maybe enough to slightly augment the action potential to reach the presynaptic terminal.

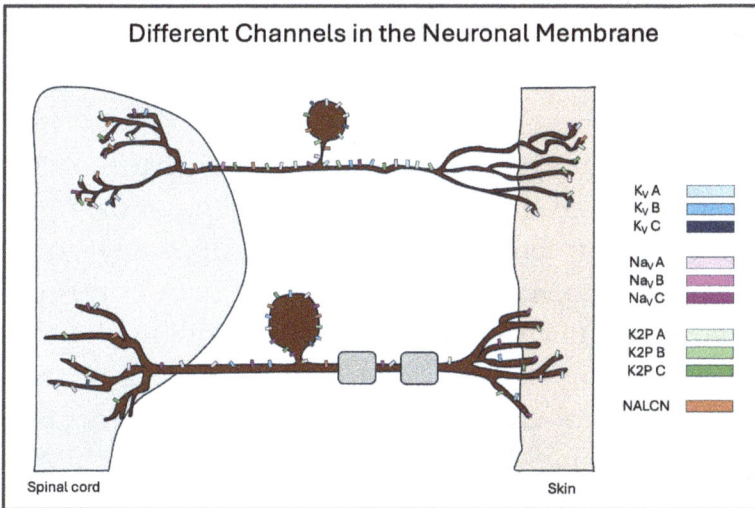

Figure 13: *A generalized figure of unmyelinated and myelinated sensory neurons relaying information from the skin to the spinal cord of a vertebrate animal. The emphasis is on the variation in the distribution and density of the subtypes in K^+_V, Na^+_V, K2P channels along with NALCN channels on the membrane of the neurons. (Modified from Tsantoulas and McMahon, 2014).*

Some sensory endings, such as those that monitor temperature, have a thermal receptor which can open as a Ca^{2+} and/or a Na^+ channel. The graded response from the sensory ending can then trigger an action potential to be induced if the threshold is met for the voltage gated-Na^+ channels, and there is a high density of the channels. The thermal response in larval *Drosophila* will be illustrated in the next chapter of this book. Other types of sensory receptors, such as mechanosensory ones, have stretch-activated ion channels in their sensory endings which can be selective to various types of ions. It is interesting to note that some subtypes of K2P are also sensitive to deformation of the membrane and might serve multiple roles as a channel helping to maintain a stable membrane potential but also acting as a monitor of membrane deformation.

In understanding more about how the membrane potential can be altered, one can predict how various pharmacological agents could impact the shape of action potentials or even the resting membrane potential to aid in therapeutic treatments. However, predictions only go so far without experimentation. For example, it was postulated that the mechanism of how 4-Aminopyridine (4-AP; also called Ampyra or dalfampridine in a slow release therapeutic form) can be beneficial for some patients with multiple sclerosis (MS) (Baker, 2013; Correale et al., 2017) in that it would block the fast-activating voltage-gated potassium (K^+_V) channels in the presynaptic nerve terminal; and thus, potentially cause opening of more and as well as prolong the opening of Ca^{2+}_V channels to enhance vesicle fusion leading to an increase in synaptic transmission (Clay, 1985). Theoretically, this makes sense (**Figures 14** and **15**); however, Baker (2013) makes a valid point that it would take about 100 microM to have much of an effect in widening the action potential at the presynaptic motor nerve terminals and that this would be a toxic level for mammals. It is likely that the medication is having an action on other cell types involved with the inflammatory responses associated with MS. MS has a complex pathology with various cell types involved, so it may be hard to pinpoint how 4-AP may be functioning overall as a therapeutic agent (Krishnan et al. 2013; Jensen et al. 2014; Filippi et al. 2020). An educational module was developed to cover the concept of 4-AP's action used for treatment of MS (Cooper et al., 2020). The point is that what might appear to be a valid approach to a clinical or experimental manipulation that may have compounding variables not accounted for when put into practice. A point to consider even in the exercise demonstrated above with raising extracellular Ca^{2+} concentration and blocking the NALCN channels. It is also noted that extracellular Ca^{2+} may also have an action in partially blocking voltage-gated Na^{+} channels (Armstrong and Cota, 1999; Mert et al., 2003).

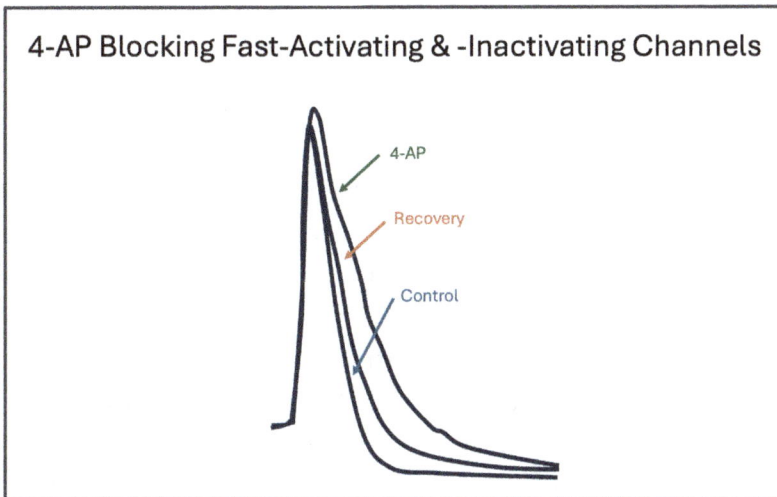

Figure 14: The general shape of an action potential in a motor neuron before, during and after exposure to 4-AP. The fast-activating K^+_V channels are inhibited by 4-AP and in theory widening the action potential in the nerve terminal.

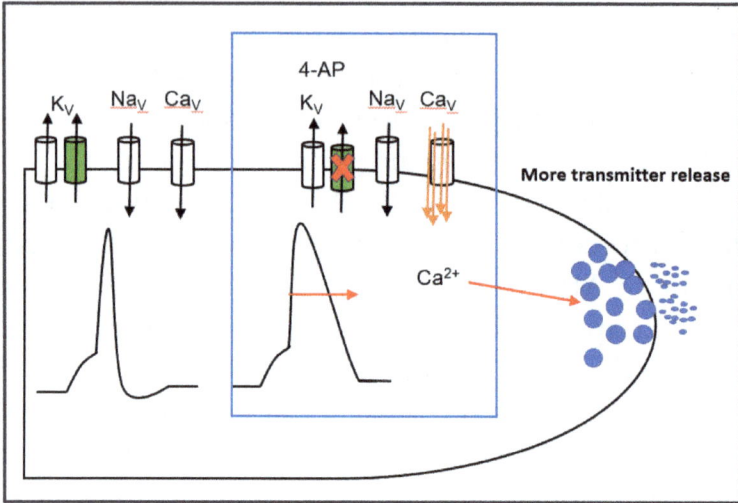

Figure 15: A schematic of the potential action of using 4-AP in blocking a subtype of K^+_V channel and the effect on enhancing Ca^{2+} entry, by keeping the Ca^{2+}_V channels open longer, in the terminal to promote more evoked synaptic vesicle fusion events (modified from Cooper et al., 2020).

Another concept that has wide application in altering neuronal excitability is on the topic of the inactivation and removal of inactivation of the voltage-gated Na^+ channel (Na^+_V). Most physiology classes cover absolute and relative refractory periods of neurons. Students in such classes have dealt with the concept that the Na^+_V undergoes inactivation during the upstroke of an action potential, and the channels remain in an inactivated state until the membrane potential comes back to a resting potential state to slowly remove the inactivation (i.e.,

44

conformational change of the channel). What is not widely mentioned in textbooks is that hyperpolarization in a neuron can remove the inactivated state of Na^+_V, which is present at the normal resting membrane potential. Thus, hyperpolarizing a neuron and then coming back to a resting state may actually initiate an action potential, as the threshold has been lowered by the removal of the inactivation. This concept is diagrammed in **Figure 16** and appears to explain phenomena observations in mammalian central neurons (Rama et al., 2015), as well as in neurons of the leech for Ca^{2+}_V channels (Olsen and Calabrese, 1996). Considering synaptic transmission which allows Cl⁻ influx or a temperature change which hyperpolarizes the membrane or an anesthetic which activates some subtypes of K2P channels, all potentially aid in removing inactivation of Na^+_V channels; thus, allowing the membrane to be more excitable upon removal of the hyperpolarization.

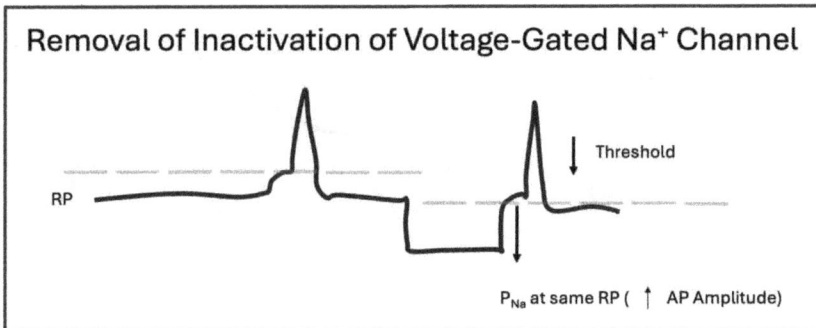

Removal of Inactivation of Voltage-Gated Na⁺ Channel

RP

Threshold

P_{Na} at same RP (↑ AP Amplitude)

Figure 16: *Removal of inactivation of the Na^+_V channels by hyperpolarization reduces the threshold of activation.*

Likewise, somewhat of an opposite scenario occurs with slow depolarization, where the threshold of activation is increased. Slow depolarization of the membrane can open a few Na^+_V channels and a few more as the depolarization continues. This slowly leads to inactivation of more and more Na^+_V channels so the membrane potential can exceed the normal threshold without activating an action potential. This concept is schematically shown in **Figure 17**. One can slowly inject current into a cell to open some Na^+_V channels and observe the differences in reaching a threshold to induce an action potential. Such experiments can be readily performed using the large neurons in the leech model (Muller et al., 1982; Titlow et al., 2013). This has practical application in pathological situations with hyperkalemia or with tissue injury leading to hyperkalemia. Educational models for experimentation have been developed over these concepts (Malloy et al., 2017; Thenappan et al., 2017; Cooper et al., 2019b).

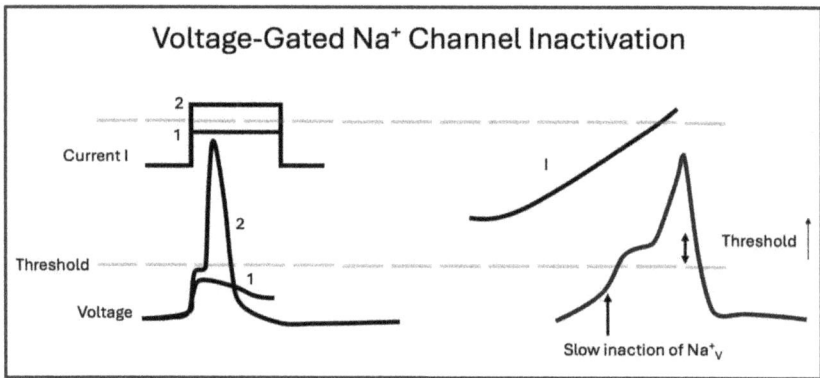

Voltage-Gated Na$^+$ Channel Inactivation

Figure 17: Illustrating the effect of slow inactivation of the Na^+_V channels by gradual depolarization. The slow depolarization inactivates a subset of the Na^+_V channels; thus, raising the threshold of initiating an action potential. On the left panel, threshold is met with a stronger depolarization above a graded response; however, on the right panel with slow current injection the threshold has now been increased ("I" represents current being injected).

Other means in which the threshold of activation may be altered among cells is the varied expression in the types of channels. A practical example is in the expression of the types of leak channels (i.e., K2P and NALCN) which maintain the resting membrane potential. If a cell or a region of a cell had a difference in the expression of K2P channels, which were constantly open, the membrane may indeed have a more negative resting membrane potential. This could lead to many differences in the shape of an action potential, such as lowering the threshold for Na^+_V if any residual inactivation was present in the condition without the over expression of K2P channels. In addition, the driving gradient to the E_{Na} would be larger; in turn, the amplitude of the action potential may be larger.

As a proof of concept, larval *Drosophila* muscles were genetically induced to overexpress a subtype of K2P channels, and the resting membrane potential was on average approximately 20 mV more negative than the control larvae for the same bathing media (Elliott and Cooper, 2024). This has yet to be examined within the motor nerve axon, as one is not able to obtain an intracellular recording in the axon of larval *Drosophila*; however, the cell bodies of some neurons in the larval brain could be recorded from with intracellular electrodes to document alterations in electrical events with cells overexpressing K2P channels. The anticipated differences may resemble differences as shown in **Figure 18**.

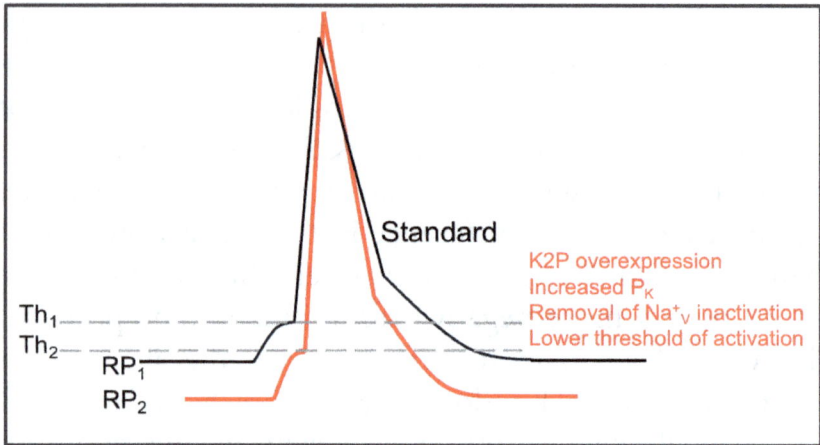

Figure 18: *Two theoretical action potentials in two different neurons. One a standard neuron and one over expressing a subtype of K2P channels which lowers the resting membrane potential from RP₁ to RP₂. Thus, the threshold of activation for inducing an action potential is more negative due to removal of any residual inactivation of the Na⁺ᵥ channels and could even result in a larger amplitude action potential.*

1.6. Factors That Affect Membrane Potential

The contribution of pumps and ion exchanges vary from cell to cell and the surrounding conditions (e.g., pH, differences in membrane potential, differences in ionic concentrations across the membrane, etc.) all can affect membrane potential. Classic experiments with injection of Na^+ and altering external K^+ concentration revealed that the pump can significantly hyperpolarize a membrane due to the greater ability to pump Na^+ out of a cell compared to K^+ into a cell as demonstrated in snail neurons (Thomas, 1969). An additional complexity is that the pump's efficiency is temperature-dependent (Fischbarg, 1972). The contribution of the Na^+-K^+ pump can be observed by blocking the pump's action with blockers or stopping ATP production to drive the pump. Blocking ATP production, of course, can affect other pumps that might be present in the membrane.

An example of using Ouabain to block the Na^+-K^+ pump in larval *Drosophila* muscle illustrates how rapidly the effect can be and to the degree the pump contributes to maintaining the membrane potential (**Figure 19**).

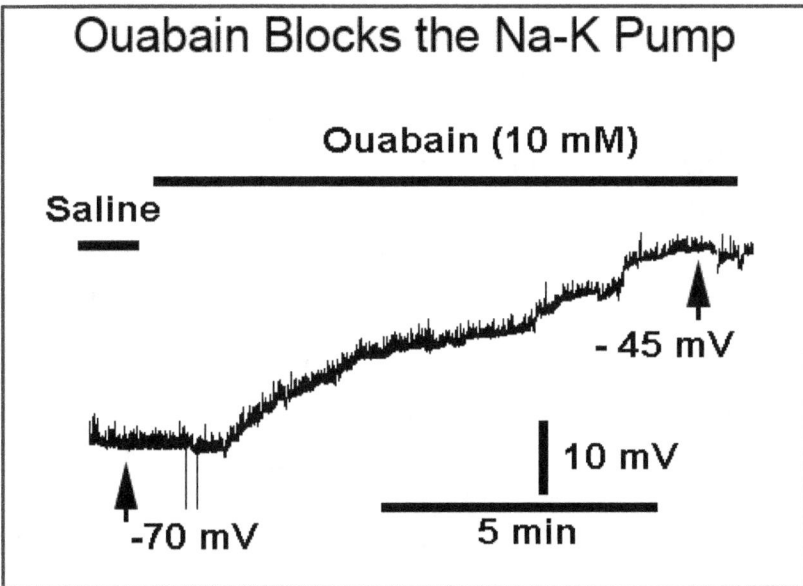

Ouabain Blocks the Na-K Pump

Ouabain (10 mM)

Saline

- 45 mV

10 mV

-70 mV

5 min

Figure 19: *The effect of blocking the Na^+-K^+ ATPase pump in larval Drosophila muscle. Within 10 minutes of incubation with Ouabain (10 mM) the membrane depolarized from -70 mV to -45 mV illustrating how much the pump is contributing to the resting membrane potential. (Figure reproduced from Potter et al., 2021).*

Other factors which are not addressed in the simulations provided herein or in many estimates of membrane potential in the primary literature, as well as general physiology textbooks, do not address the effects that **exchangers**, such as the sodium calcium exchanger (i.e. NCX), have on the resting membrane potential or for that matter in the shape of an action potential when the flux of Na^+ and Ca^{2+} are substantially influencing the NCX activity. It has been estimated for the NCX that for every Ca^{2+} ion that moves across the membrane there is a match of three Na^+ ions (Caputo, et al., 1989; see Nicholls et al., 1992, pp-86-87). Similarly, Cl^- can also be exchanged for HCO_3^- and Na^+ for H^+ depending on the conditions. These exchanges come more into play in aiding maintaining pH balance, which can be substantial during cellular activity when the production of CO_2 is increased and subsequently producing HCO_3^- and H^+.

An example of the impact of an altered pH environment is membrane potential monitored in larval *Drosophila* muscle, while changing the saline with a pH of 7.2 to 5.0 (Badre et al., 2005). The resting membrane potential changed from around -60 mV to -40 mV within 2 minutes of exposure (**Figure 20**). It was also noted that bubbling the physiological saline with 100% CO_2 for 10 minutes reduced the pH from 7.2 to 5.0. This was one reason to examine the effects of adjusting pH with HCl to 5.0 and examining the effect on the membrane potential to address the effect of CO_2.

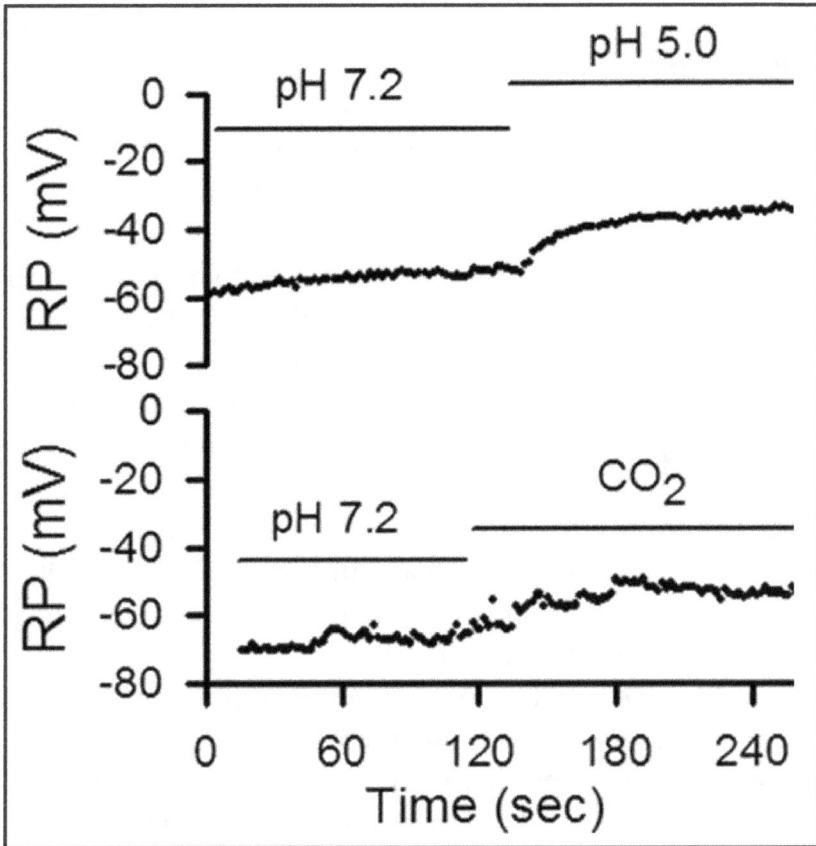

Figure 20: *The effect of pH on membrane potential. The larval muscle of Drosophila is very sensitive to changes in pH of the bathing saline. The effects of bubbled CO_2 in saline or added HCl to the saline to bring the pH from 7.2 to 5.0 depolarized the membrane (Modified from Badre et al., 2005).*

The depolarization of the membrane by acidic pH on the resting membrane potential is likely the result of some subtypes of K2P channels being inhibited (Kamuene et al., 2021; Cunningham et al., 2020). The ability to inhibit K2P channels through pharmacological actions (i.e. Doxapram hydrochloride marketed as Dopram, Stimulex or Respiram) has aided in the ability to therapeutically drive respiration in intact mammals by tricking the chemoreception in thinking the body is acidic. This provides a therapeutic drive for one to ventilate on their own with decreased use of a mechanical ventilator for humans when in a comprised health condition. Such treatments are used during the induction of therapeutic hypothermia and was used during the era of COVID-19 (Sanders et al. 2020; Baxter 1976; Kim 2005; Fathi et al. 2020; Cotten et al. 2006)

As a proof of concept of doxapram directly depolarizing cells as would in an acidic environment, the compound was tested on crayfish and *Drosophila* muscle fibers and motor neurons of the crayfish while monitoring the membrane potential (Vacassenno et al., 2023a,b; Elliott et al., 2024). The compound showed a dose-dependent effect in depolarizing the membrane potentials.

Oddly enough, another interesting phenomenon which impacts the resting membrane of cells is the **lipopolysaccharides** (i.e. LPS), also known as endotoxin, which is a structural component from some forms of Gram-negative bacteria. Few studies have examined the direct effects of LPS on membrane potential independent of an immune response by cytokines (Cooper et al., 2019a; Cooper and Krall, 2022). Interestingly, the muscle of larval *Drosophila* showed enhanced effects to LPS, compared to muscles of frog or crayfish (Ballinger-Boone et al., 2020). The rapid, acute response on *Drosophila* muscle within a second is a hyperpolarization for a minute, followed by a gradual depolarization. It appeared that the effects were mediated via an action on a subtype of K2P channels and not hyperactivation of the Na^+-K^+ ATPase pump or a Cl^- influx (Potter et al., 2021), so the next logical step was to determine if doxapram could block the response of LPS.

Turns out, doxapram does block LPs response in both the *Drosophila* and crayfish models (Brock and Cooper, 2023; Elliott et al., 2024).

By understanding the concepts that drive the membrane potential, one can predict the mechanisms of a phenomenon and test them. This is the beauty of the scientific process. Recall, the action above mentioned by LPS resulted in an acute hyperpolarization, followed by depolarization. It appears the hyperpolarization was mechanistically examined and even tested by over-expressing the K2P channels in *Drosophila* muscle to see how the muscle would respond to LPS (Hadjisavva and Cooper, 2025). The muscle hyperpolarized even more than just to the overexpression of K2P channels to as much as -100 mV! However, the effect on depolarization has not been resolved. It was examined and determined that there was no effect on blocking the Na^+-K^+ ATPase pump or a result of action on Na^+_V channels or delayed blocking of K2P channels, but it appears as a delay in activating NALCN leak channels. This was examined by substituting and eliminating the Na^+ in the external bathing environment as well as removing external K^+ driving the potential toward a E_K in control muscle, as well as in muscles overexpressing K2P channels (Hadjisavva, and Cooper, 2025). The prolonged hyperpolarization only occurred if Na^+ was not present in the bathing media (**Figure 21**).

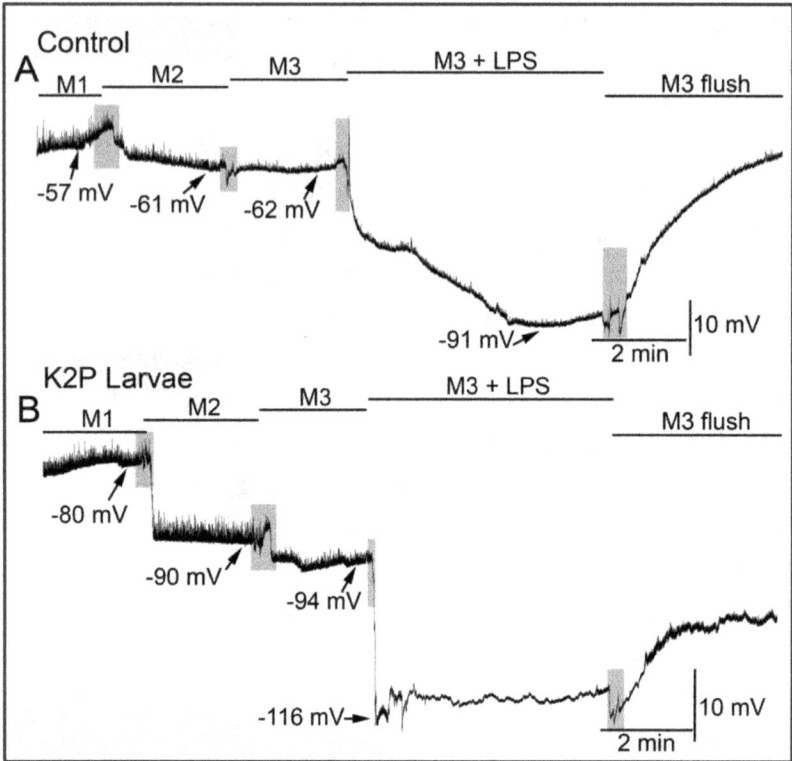

Figure 21: *The effect of LPS on membrane potential with the saline where KCl and NaCl were kept to a minimal concentration (Figure 6 taken from Hadjisavva, and Cooper, 2025). Control larvae (A) and larvae overexpressing K2P channels (B) in muscles of larval Drosophila. For details on the bathing solutions used (M1, M2, and M3) see primary manuscript Hadjisavva, and Cooper, 2025.*

In Chapter 2, the behavior of larval *Drosophila* is used as a model to demonstrate the effect of temperature. In a cold environment, the larvae are rapidly attracted to the warm environment in the testing arena used. Larvae can sense temperature by receptors in the sensory neurons. The distribution and density of these receptors could also vary among larvae, allowing some to be more or less sensitive to temperature. In *Drosophila*, a receptor termed TRPA1 senses heat and can itself become an ion channel allowing the influx of Ca^{2+} as well as Na^+ into the cell (Luo et al., 2017). Thus, the membrane potential will depolarize. So, the effect of temperature on membrane potential is wide ranging from effects on the equilibrium potential for each ion, on pumps, exchangers, and that which detect temperature, making it difficult to account for all the changing variables in simulations with changes in temperature.

Critical Thinking
& Thought Questions

1. How does temperature affect membrane potential?

2. How does temperature affect the shape of an action potential?

3. How does extracellular Ca^{2+} potentially influence the resting membrane potential?

4. How extracellular Ca^{2+} potentially influence the shape of an action potential?

5. If there appears to be various pharmacological compounds that block different types of K^+_V, how could these selective blockers have an influence on the shape of an action potential?

6. Why is it so difficult to computationally describe the electrical properties of cells in different environments, such as temperature or with pharmacological agents?

7. When one is asked to draw a standardized shape of an action potential what would one draw? What should one add in a disclaimer about the drawing?

8. Why does a single cell appear to have different shapes of electrical events in different regions?

9. Take a look at the figure below. In a recording of the membrane potential of a larval *Drosophila* muscle, warm

saline was added (where the large downward deflection in the trace is seen) and then allowed to cool back to room temperature in a minute. Why did this happen?

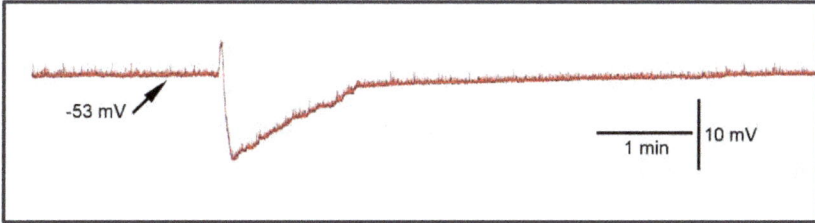

10. How might you respond if this figure below was shown to you and say the same experiment was conducted as above (i.e., warm saline was added where the large upward deflection in the trace is seen) but on a genetically modified larval *Drosophila*? What might you postulate that is different between these two *Drosophila* preparations?

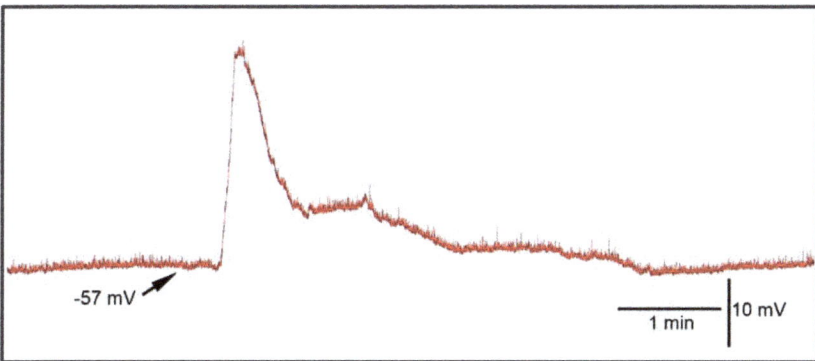

11. What other factors impact membrane potential? The shape of an action potential?

12. Which factor (e.g., ion concentration, ion channel, temperature, etc.) do you think plays the most significant

role in predicting the membrane potential? The shape of an action potential?

13. In most cells, why is the resting membrane potential negative?

14. Why does K2P channel overexpression cause hyperpolarization?

15. Why does TRPA1 channel (Ca^{2+}-permeable channel) overexpression cause depolarization?

16. What are some advantages of using Goldman-Hodgkin-Katz simulations to predict action potential, particularly when compared to the standard equation? Any disadvantages?

References

Armstrong, C. M., Cota, G. 1999. Calcium block of Na^+ channels and its effect on closing rate.
Proceedings of the National Academy of Sciences of the United States of America. 96(7):4154–4157.
https://doi.org/10.1073/pnas.96.7.4154.

Atwood, H. L. 1982. Chapter 9: Membrane Physiology of Invertebrates (ed. R. B. Podesta).
New York: Marcel Dekker. Pp: 341-407.
ISBN 10: 0824715039, ISBN 13: 9780824715038.

Baierlein, B., Thurow, A.L., Atwood, H. L., Cooper, R.L. 2011. Membrane potentials, synaptic responses, neuronal circuitry, neuromodulation and muscle histology using the crayfish: student laboratory exercises.
Journal of Visualized Experiments: JoVE. (47): 2322.
https://doi.org/10.3791/2322.

Badre, N.H., Martin, M.E., Cooper, R.L. 2005. The physiological and behavioral effects of carbon dioxide on *Drosophila* larvae.
Comparative Biochemistry and Physiology A. 140:363-376.

Baker, M.D. 2013. Potential therapeutic mechanism of K^+ channel block for MS.
Multiple Sclerosis and Related Disorders. 2(4): 270-280.
https://doi.org/10.1016/j.msard.2013.01.005.

Ballinger-Boone, C., Anyagaligbo, O., Bernard, J., Bierbower, S.M., Dupont-Versteegden, E.E., Ghoweri, A., Greenhalgh, A., Harrison, D., Istas, O., McNabb, M., et al. 2020. The effects of bacterial endotoxin (LPS) on cardiac and synaptic function in various animal models: Larval *Drosophila*, crayfish, crab, and rodent.
International Journal of Zoological Research. 16:33–62.
doi: 10.3923/ijzr.2020.33.62.

Baxter, A.D. 1976. Side effects of doxapram infusion. Eur. J. Intensive Care Med. 2(2):87–88.
doi.org/10.1007/BF01886121.

Brock, K.E, Cooper, R.L. 2023. The effects of doxapram blocking the response of Gram-negative bacterial toxin (LPS) at glutamatergic synapses.
BIOLOGY 2023. 12(8):1046.
https://www.mdpi.com/2079-7737/12/8/1046.

Bruus, K., Kristensen, P., Larsen, E.H. 1976. Pathways for chloride and sodium transport across toad skin.
Acta Physiol. Scand. 97(1):31-47.
doi: 10.1111/j.1748-1716.1976.tb10233.x.

Caputo, C., Bezanilla, F., DiPolo, R. 1989. Currents related to the sodium-calcium exchange in squid giant axon.
Biochim. Biophys. Acta. 986(2):250-256.
doi: 10.1016/0005-2736(89)90474-4.

Cheslock, A., Andersen, M.K., MacMillan, H.A. 2021. Thermal acclimation alters Na^+/K^+-ATPase activity in a tissue-specific manner in Drosophila melanogaster. Comparative Biochemistry and Physiology.

Part A, Molecular & integrative Physiology. 256:110934.
https://doi.org/10.1016/j.cbpa.2021.110934.

Clay, J.R. 1985. Potassium current in the squid giant axon.
International Review of Neurobiology. 27:363–384.
https://doi.org/10.1016/s0074-7742(08)60562-0.

Cochet-Bissuel, M., Lory, P., Monteil, A. 2014. The sodium
leak channel, NALCN, in health and disease.
Frontiers in Cellular Neuroscience. 8:132.
https://doi.org/10.3389/fncel.2014.00132.

Cooper, R.L., Krall, R.M. 2022. Hyperpolarization induced
by LPS, but not by chloroform, is inhibited by Doxapram,
an inhibitor of two-P-domain K^+ channel (K2P).
International Journal of Molecular Sciences. 23(24):15787.
https://doi.org/10.3390/ijms232415787.

Cooper, R.L., Krall, R.M., Schultz, M.P., O'Neil, A.S.,
Dupont-Versteegden, E.E. 2020. Educational modules of
skeletal muscle anatomy and function with models and
active data gathering related to muscular dystrophy. Article
64 In: McMahon K, editor.
Advances in biology laboratory education. Volume 41.
Publication of the 41st Conference of the Association for
Biology Laboratory Education (ABLE).
https://doi.org/10.37590/able.v41.art64.

Cooper, R.L., McNabb, M., Nadolski, J. 2019a. The effects
of a bacterial endotoxin LPS on synaptic transmission at
the neuromuscular junction.
Heliyon-Elsevier. 5 (2019):e01430.
https://www.heliyon.com/article/e01430.

Cooper, R.L. Thenappan, A., Dupont-Versteegden, E.E. 2019b. Examining motor and sensory units as an educational model for understanding the impact of localized tissue injury on healthy cells. Article 6. In: McMahon K, editor. Tested studies for laboratory teaching. Volume 40. Proceedings of the 39th Conference of the Association for Biology Laboratory Education (ABLE).

Correale, J., Gaitán, M.I., Ysrraelit, M.C., Fiol, M.P. 2017. Progressive multiple sclerosis: from pathogenic mechanisms to treatment.
Brain: A Journal of Neurology. 140(3):527–546.
https://doi.org/10.1093/brain/aww258.

Cotton, J.F., Keshavaprasad. B., Laster, M.J., Eger, E.I., Yost. C.S. 2006 The ventilatory stimulant doxapram inhibits TASK tandem pore (K2P) potassium channel function but does not affect minimum alveolar anesthetic concentration.
Anesth. Analg. 102(3):779–785.
https://doi.org/10.1213/01.ane.0000194289.3.

Cunningham, K.P., MacIntyre, D.E., Mathie, A., Veale, E.L. 2020. Effects of the ventilatory stimulant, doxapram on human TASK-3 (KCNK9, K2P9.1) channels and TASK-1 (KCNK3, K2P3.1) channels.
Acta Physiol. (Oxford) 228(2):e13361.
https://doi.org/10.1111/apha.13361.

de Castro, M.A., Kim, Y., Kim, J., Cooper, R.L. 2025. Comparisons in the responses to reduced extracellular Ca^{2+} on membrane potential and hyperexcitability of neurons among various model preparations. Society for

Neuroscience Annual Meeting. San Diego, California, USA. Nov. 15-19 (Abstract presented and published).

Desai-Shah, M., Cooper, R.L. 2010. Actions of NCX, PMCA and SERCA on short-term facilitation and maintenance of transmission in nerve terminals.
The Open Physiology Journal. 3:37-50.
doi: 10.2174/1874360901003010037.

Elliott, E.R., Brock, K.E., Vacassenno, R.M., Harrison, D.A., Cooper, R.L. 2024. The effects of doxapram and its potential interactions with K2P channels in experimental model preparations.
Journal of Comparative Physiology A. 210:869–884.
https://doi.org/10.1007/s00359-024-01705-6.

Elliott, E.R., Cooper, R.L. 2024. The effect of calcium ions on resting membrane potential.
Biology. 13(9):750.
https://doi.org/10.3390/biology13090750.

Fathi, M., Massoudi, N., Nooraee, N., Beheshti Monfared, R. 2020. The effects of doxapram on time to tracheal extubation and early recovery in young morbidly obese patients scheduled for bariatric surgery: a randomised controlled trial.
Eur. J. Anaesthesiol. 37(6):457–465.
https://doi.org/10.1097/EJA.0000000000001144.

Feliciangeli, S., Chatelain, F. C., Bichet, D., Lesage, F. 2015. The family of K2P channels: salient structural and functional properties.
The Journal of Physiology. 593(12):2587–2603.
https://doi.org/10.1113/jphysiol.2014.287268.

Fettiplace, R., Fuchs, P.A. 1999, Mechanisms of hair cell tuning.
Annual Rev. Physiol. 61:809-834.
doi: 10.1146/annurev.physiol.61.1.809.

Filippi, M., Preziosa, P., Langdon, D., Lassmann, H., Paul, F., Rovira, À., Schoonheim, M.M., Solari, A., Stankoff, B., Rocca, M.A. 2020. Identifying progression in multiple sclerosis: New perspectives.
Annals of Neurology. 88(3):438–452.
https://doi.org/10.1002/ana.25808.

Fischbarg, J. 1972. Ionic permeability changes as the basis of the thermal dependence of the resting potential in barnacle muscle fibres.
The Journal of Physiology. 224(1):149–171.
https://doi.org/10.1113/jphysiol.1972.sp009886.

Gardner, C.L., Nonner, W., Eisenberg, R.S. 2004. Electrodiffusion model simulation of ionic channels: 1D simulations.
Journal of Computational Electronics. 3:25-31.
https://doi.org/10.1023/B:JCEL.0000029453.09980.fb.

Goldman, D.E. 1943. Potential, impedance, and rectification in membranes.
The Journal of General Physiology. 27(1):37–60.
https://doi.org/10.1085/jgp.27.1.37

Goldstein, S.A., Wang, K.W., Ilan, N., Pausch, M.H. 1998. Sequence and function of the two P domain potassium channels: implications of an emerging superfamily.
Journal of Molecular Medicine (Berlin, Germany). 76(1):13–20.
https://doi.org/10.1007/s001090050186.

Hadjisavva, M.E., Cooper, R.L. 2025. The biphasic effect of lipopolysaccharide on membrane potential.
Membranes 2025. 15:74.
https://doi.org/10.3390/membranes15030074.

Hernandez, J., Fischbarg, J., Liebovitch, L.S. 1989. Kinetic model of the effects of electrogenic enzymes on the membrane potential.
Journal of Theoretical Biology. 137(1):113–125.
https://doi.org/10.1016/s0022-5193(89)80153-5.

Hille, B. 1992. Ionic Channels of Excitable Membranes, 2nd ed., Sinauer Assoc., Sunderland, Mass. USA.
ISBN-13 978-0878933211.

Hodgkin, A.L., Huxley, A.F. 1952. A quantitative description of membrane current and its application to conduction and excitation in nerve.
The Journal of Physiology. 117(4):500–544.
https://doi.org/10.1113/jphysiol.1952.sp004764.

Hodgkin, A.L., Huxley, A.F., KATZ, B. 1952. Measurement of current-voltage relations in the membrane of the giant axon of Loligo.
The Journal of Physiology. 116(4):424–448.
https://doi.org/10.1113/jphysiol.1952.sp004716.

Hodgkin, A.L., Katz, B. 1949. The effect of sodium ions on the electrical activity of giant axon of the squid.
The Journal of Physiology. 108(1):37–77.
https://doi.org/10.1113/jphysiol.1949.sp004310.

Jensen, H.B., Ravnborg, M., Dalgas, U., Stenager, E. 2014. 4-Aminopyridine for symptomatic treatment of multiple sclerosis: a systematic review.
Therapeutic Advances in Neurological Disorders. 7(2):97–113.
https://doi.org/10.1177/1756285613512712.

Jørgensen, C.B., Levi, H., Zerahn, K. 1954. On active uptake of sodium and chloride ions in anurans.
Acta Physiol. Scand. 30(2-3):178-190.

Kamuene, J.M., Xu, Y., Plant, L.D. 2021. The pharmacology of two-pore domain potassium channels.
Handbook of Experimental Pharmacology. 267:417–443.
https://doi.org/10.1007/164_2021_462.

Katz, G.M., Reuben, J.P., MacBerman et al. 1972. Potassium redistribution and water movement in crayfish muscle fibers.
J. Comp. Physiol. 80:267–283.
https://doi.org/10.1007/BF00694841.

Kim, D. 2005. Physiology and pharmacology of two-pore domain potassium channels.
Curr. Pharm. Des. 11(21):2717–2736.
https://doi.org/10.2174/1381612054546824.

Krans, J.L., Parfitt, K.D., Gawera, K.D., Rivlin, P.K., Hoy, R.R. 2010. The resting membrane potential of *Drosophila melanogaster* larval muscle depends strongly on external calcium concentration.
Journal of Insect Physiology. 56(3):304–313.
https://doi.org/10.1016/j.jinsphys.2009.11.002.

Krishnan, A.V., Kiernan, M.C. 2013. Sustained-release fampridine and the role of ion channel dysfunction in multiple sclerosis.
Multiple Sclerosis (Houndmills, Basingstoke, England). 19(4):385–391.
https://doi.org/10.1177/1352458512463769.

Krogh, A. 1904. On the cutaneous and pulmonary respiration of the frog. A contribution to the theory of the gas exchange between the blood and atmosphere.
Skand. Arch. Physiol. 15:328-419.

Krogh, A. 1937. Osmotic regulation in the frog *(R. esculenta)* by active absorption of chloride ions.
Skand. Arch. Physiol. 76:60-74.

Krogh, A. 1938. The active absorption of ions in some freshwater animals. Z. vergl.
Physiol. 25:335-350.

Lee, J.H., Cribbs, L.L., Perez-Reyes, E. 1999. Cloning of a novel four repeat protein related to voltage-gated sodium and calcium channels.
FEBS Letters. 445(2-3):231–236.
https://doi.org/10.1016/s0014-5793(99)00082-4.

Lin, J,W. 2016. Na^+ current in presynaptic terminals of the crayfish opener cannot initiate action potentials.
J. Neuro-physiol. 115(1):617-621.
doi: 10.1152/jn.00959.2015.

Ling, G.N. 1992. A Revolution in the Physiology of the Living Cell. Krieger Publishing Co.; Malabar, FL, USA.

Liu, W. 2009. One-dimensional steady-state Poisson-Nernst-Planck systems for ion channels with multiple ion species.
Journal of Differential Equations. 246:428451.
https://doi.org/10.1016/j.jde.2008.09.010.

Lu, B., Su, Y., Das, S., Liu, J., Xia, J., Ren, D. 2007. The neuronal channel NALCN contributes resting sodium permeability and is required for normal respiratory rhythm.
Cell. 129(2):371–383.
https://doi.org/10.1016/j.cell.2007.02.041.

Luo, J., Shen, W.L., Montell, C. 2017. TRPA1 mediates sensation of the rate of temperature change in *Drosophila* larvae.
Nature Neuroscience. 20(1):34–41.
https://doi.org/10.1038/nn.4416.

Malloy, C., Dayaram,V., Martha, S., Alvarez, B., Chukwudolue, I., Dabbain, N., et al., 2017. The effects of potassium and muscle homogenate on proprioceptive responses in crayfish and crab.
J. of Exp. Zoology. 327(6):366–379.

Matthews, G. 1996. Neurotransmitter release. Annual Review of Neuroscience. 19:219–233.
https://doi.org/10.1146/annurev.ne.19.030196.001251.

Mert, T., Gunes, Y., Guven, M., Gunay, I., & Ozcengiz, D. 2003. Effects of calcium and magnesium on peripheral nerve conduction.
Polish Journal of Pharmacology. 55(1):25–30.

Monteil, A., Guérineau, N. C., Gil-Nagel, A., Parra-Diaz, P., Lory, P., Senatore, A. 2024. New insights into the physiology and pathophysiology of the atypical sodium leak channel NALCN.
Physiological Reviews. 104(1): 399–472.
https://doi.org/10.1152/physrev.00014.2022.

Mullins, L.J., Noda, K. 1963. The influence of sodium-free solutions on the membrane potential of frog muscle fibers.
J. Gen. Physiol. 47(1):117-132.
doi: 10.1085/Jgp.47.1.117.

Muller, K.J., Nicholls, J.G., Stent, G.S. 1981. Neurobiology of the Leech. Eds. Kenneth J. Muller, John G. Nicholls, and Gunther S. Stent. Cold Spring Harbor Laboratory, Cold Spring Harbor, N.Y., 1981. 2nd ed.

Nernst, W.H. 1888. Zur Kinetik der Lösung befindlichen Körper: Theorie der Diffusion. Z. Phys. Chem. 3: 613-637.

Nernst, W.H. 1889. Die elektromotorische Wirksamkeit der Ionen. Z. Phys. Chem. 4:129-181.

Nicholls, J.G. Martin, A.R., Wallace, B.G. 1992. From Neuron to Brain. A Cellular and Molecular Approach to the Function of the Nervous System. 3rd ed. Eds. John G. Nicholls, A. Robert Martin, Bruce G. Wallace. Sinauer Associates, Inc. Sunderland, Mass. USA.

Olsen, O.H., Calabrese, R.L. 1996. Activation of intrinsic and synaptic currents in leech heart interneurons by realistic waveforms. J. Neurosci. 16(16):4958-4970. doi:10.1523/JNEUROSCI.16-16-04958.1996.

Plant, L.D., Goldstein, S.A.N. 2015. Two-pore domain potassium channels. In Handbook of Ion Channels, 1st ed.; Zheng, J., Trudeau, M.C., Eds.; CRC Press: Boca Raton, FL, USA, pp. 261–274. ISBN 9780429193965.

Potter, R., Meade, A., Potter, S., Cooper, R.L. 2021. Rapid and direct action of lipopolysaccharide (LPS) on skeletal muscle of larval *Drosophila*. Biology. 10:1235. https://doi.org/10.3390/biology10121235.

Prelic. S., Pal Mahadevan, V., Venkateswaran, V., Lavista-Llanos, S., Hansson, B.S., Wicher, D. 2022. Functional interaction between *Drosophila* olfactory sensory neurons and their support cells.
Front Cell Neurosci. 15:789086.
doi: 10.3389/fncel.2021.789086.

Rama, S., Zbili, M., Bialowas, A., Fronzaroli-Molinieres, L., Ankri, N., Carlier, E., Marra, V., Debanne, D. 2015. Presynaptic hyperpolarization induces a fast analogue modulation of spike-evoked transmission mediated by axonal sodium channels.
Nature Communications. 6:10163.
https://doi.org/10.1038/ncomms10163.

Ruff, R.L. 1999. Effects of temperature on slow and fast inactivation of rat skeletal muscle Na^+ channels. The American Journal of Physiology. 277(5):C937–C947.
https://doi.org/10.1152/ajpcell.1999.277.5.C937.

Sanders, J.M., Monogue, M.L., Jodlowski, T.Z., Cutrell, J.B. 2020. Pharmacologic treatments for coronavirus disease 2019 (COVID-19). Review.
JAMA 323(18):1824–1836.
https://doi.org/10.1001/jama.2020.6019.

Skou, J.C. 1965. Enzymatic basis for active transport of Na^+ and K^+ across cell membrane.
Physiological Reviews. 45:596–617.
https://doi.org/10.1152/physrev.1965.45.3.596.

Skou J.C. 1998. Nobel Lecture. The identification of the sodium pump.
Bioscience reports, 18(4):155–169.
https://doi.org/10.1023/a:1020196612909

Skou, J.C. 1989a. The influence of some cations on an adenosine triphosphatase from peripheral nerves.
Biochim. Biophys. Acta. 1000:439–446.

Skou, J.C. 1989b. The identification of the sodium-pump as the membrane-bound Na^+/K^+-ATPase: a commentary on 'The Influence of Some Cations on an Adenosine Triphosphatase from Peripheral Nerves'.
Biochim. Biophys. Acta. 1000:435–438.

Tamagawa, H., Morita, S. 2014. Membrane potential generated by ion adsorption.
Membranes. 4(2):257–274.
https://doi.org/10.3390/membranes4020257.

Thenappan, A., Dupont-Versteegden, E.E., Cooper, R.L. 2019. An educational model for understanding acute deep tissue injury of motor units.
Journal of Young Investigators. 36(5):62-71.

Thomas, R.C. 1969. Membrane current and intracellular sodium changes in a snail neuron during extrusion of injected sodium.
J. Physiol. 201(2):495-514.
doi: 10.1113/jphysiol.1969.sp008769.

Thomas, R.C. 1977. The role of bicarbonate, chloride and sodium ions in the regulation of intracellular pH in snail neurones.
J. Physiol. 273(1):317-338.
doi: 10.1113/jphysiol.1977.sp012096.

Titlow, J., Majeed, Z.R., Nicholls, J.G., Cooper, R.L. 2013. Intracellular recording, sensory field mapping, and culturing identified neurons in the leech, Hirudo medicinalis.
Journal of Visualized Experiments (JoVE). 81:e50631.
doi:10.3791/50631.

Tsantoulas, C., McMahon, S.B. 2014. Opening paths to novel analgesics: the role of potassium channels in chronic pain.
Trends in Neurosciences. 37(3):146–158.
https://doi.org/10.1016/j.tins.2013.12.002.

Vacassenno, R.M., Haddad, C.N., Cooper, R.L. 2023a. The effects on resting membrane potential and synaptic transmission by Doxapram (blocker of K2p channels) at the *Drosophila* neuromuscular junction.
Comparative Biochemistry and Physiology Part C: Toxicology & Pharmacology. 263 (2023):109497.

Vacassenno, R.M., Haddad, C.N., Cooper, R.L. 2023b. Bacterial lipopolysaccharide hyperpolarizes the membrane potential and is antagonized by the K2p channel blocker doxapram. Comparative Biochemistry and Physiology Part C:
Toxicology & Pharmacology. 266:109571.
https://doi.org/10.1016/j.cbpc.2023.109571.

Xie, D., Lu, B. 2020. An effective finite element iterative solver for a Poisson--Nernst--Planck ion channel model with periodic boundary conditions.
SIAM Journal on Scientific Computing. 42 (6):B1490B1516.
doi: 10.1137/19M1297099.

Chapter 2:

Temperature & Electric Potential
in Invertebrate Animal Behavior
Elementary to College Level

This is an educational module that will allow students to investigate invertebrate animal behavior in environments related to temperature and electric potential. The activity presented integrates STEM concepts in biology, chemistry, math, physics, and engineering at various levels. With the use of inexpensive design and instruments, as well as readily obtained invertebrate animals, students will be able to construct mechanistic explanations of the phenomenon, as well as make and test predictions. The concepts relate to real-world application and potential solutions to practical issues.

The activities presented do not intend to harm the animal models. The objectives are to examine the effect of heat and/or electric potential on the attraction and repulsion of *Drosophila melanogaster* larvae in various conductive media with creative designing and testing various environmental arenas. The activity allows students to examine the conductive nature of a media, make measurements of electric potential and relate findings to the response of various animal models.

The activities cover an array of the next generation science standards (NGSS) and can be modified from year to year and to the academic level of the investigators. The supplies can be replicated for individuals or used as a single demonstration at a minimal cost per set-up and reused.

1.0 Introduction

Using applicable real-world problems as a theme for classroom activities are likely to be more engaging for students than teaching just the content in isolation (Marx et al., 1997; Krajcik & Czerniak, 2014). Developing a concept in a model to examine and manipulate allows the topic to be developed in a classroom or at home in controlled conditions to potentially later be scaled to a larger arena and within an applicable setting (Cole et al. 2005).

Engaging students with hands-on activities where students are collecting and analyzing data to test predictions based on chemistry, physics, mathematics, and biology allows the students to integrate these disciplines without personally knowing they are utilizing them. The need to learn what is required for the explanations of observations and develop perturbations of an experimental design for making new predictions will lead one to investigating various disciplines as needed. Guided learning in promoting curiosity about content inspires motivation to learn and improves understanding (National Research Council. 2005).

In keeping track of the concepts being taught in these modules presented one will note they cover many aspects of the Next Generation Science Standards (NGSS Lead States, 2013; NGSS-https://www.nextgenscience.org/). Focusing on solving a real-world problem may be one approach an instructor could use to introduce these modules to a classroom. Such an approach utilizes one's desire to solve problems (Norman and Schmidt, 1992).

Details of Chapter 2 are shown in the following video link: https://youtu.be/DdzR9GPOwWg

2.0 Two Research Modules

It is well established that animals can sense electric fields, from sharks assessing electrical signals in the sand underwater, to species of fish that can not only generate electric currents but also perceive them (Crampton, 2019). It is even noted that insects can sense electric fields within the air as a medium (England et al., 2024).

Invertebrates such as earthworms in soil can respond to electric fields through the soil. One common approach to collect earthworms is to put electric poles in the ground and provide a direct current. They are attracted to the negative pole of a direct current source. The abilities of animals to detect electric fields is still an area of scientific investigation. Electric fields can also generate a magnetic field and depending on the environment, a potential thermal gradient. So, the physical nature of these environments may be combined in the sensing ability of an animal to their surroundings. There are reports detailing how larval *Drosophila*, commonly known as fruit flies, have receptors to detect an electric field (Tadres et al., 2025).

The *Drosophila* genome is easily manipulated, which allows researchers to remove the genes that code for the production of various receptor proteins, resulting in an altered ability of larvae to respond to electric fields. There are many practical applications to direct animals to be attracted or repulsed to particular environments. For example, it is known that ants tend to be attracted to circuit breaker boxes on the outside of buildings. Such attraction may be due to the electric field that is generated or possibly as a thermal response or a combination of the factors. It has also been shown that some species of termites are attracted to electric circuit panels. Thus, one might be able to attract insects away from commercial food products or structures to a collection device so as to remove the animal from particular environments.

Such easy and readily accessible organisms and materials for experimentation as an educational experience for elementary school to college level students provides a model in which many questions can be asked and tested. This allows students to be true investigators in answering their proposed questions by using invertebrate models with fewer regulatory constraints, rather than using higher-level vertebrate animals. In addition, invertebrate fly larvae, earthworms, meal worms, and other similar types of organisms are readily accessible in most places for easy access at relatively low cost. Large numbers of invertebrates, such as fly larvae, can be utilized to examine group behavior with the experimental protocols presented. Also, the protocols can be varied relatively easily to test different variables of interest by the experimenter.

We used an educational prototype to emphasize the ease in setting up the material, the ability to collect data and analyze the data to make future predictions in variations of experimental procedures. The level of experimentation can be very simple to complex. The ability to construct or design one's own apparatus and environmental conditions, such as the solution used to conduct electricity can be quite involved if one wishes to emphasize the chemistry and the physics behind the experimentation. The biological aspects itself can be presented in a simple format or more detailed format and relate the topic to other animals that can sense electrical fields so as to promote investigation into scientific literature, conversation and reporting by students. The topic can also be developed to conduct novel research projects for presentations at scientific meetings.

These exercises are designed in steps to help teachers and students to develop a prototype that can be readily modified for one's needs. Please use the ideas presented, figures and protocols freely for teaching purposes.

A second module deals with the effects of temperature on attraction and repulsion by invertebrate models. The design presented is one meant to be readily developed for an elementary school class to college level laboratory with very simple and readily accessible materials. Again, this exercise covers a wide range of topics where an instructor can present it in a simple format to a very detailed description in engineering design, chemistry, and physics to biological principles. There are various aspects that can be discussed by participants, to propose experiments and test hypotheses.

There are many potential experimentation designs that can be manipulated such as raising the animals in different temperatures and then testing them in a particular temperature to effects of diet and even pollutants to the animal's preference in a thermal gradient.

One can also examine different developmental conditions, such as factors that alter the animal's ability to sense the temperature and then to see how the animal responds to a choice in temperature. For example, one can compare temperature preferences across different developmental stages or across different invertebrate species. Potential differences in diet, such as giving the animals a modulator (i.e., serotonin or dopamine) that could affect their hormone levels, may alter sensitivity to environmental temperatures. One may then be able to alter the animal's sensitivity to temperature.

There are a number of studies using *Drosophila* larvae to examine the effects of temperature on behavior. One can investigate the biology from genetics to the types of sensory neurons that sense temperature as a research project. Thus, students are able to investigate scientific literature and replicate or design their own experiments by being creative and potentially making novel discoveries.

Depending on the organism, genetic manipulations can be introduced to alter sensitivity to thermal stimuli, such as by the over- or under expression of TRPA1 protein in all or subsets of sensory neurons (Barbagallo and Garrity, 2015; Kang et al., 2011). One can monitor differences in the behaviors, quantify them, and apply statistical analysis. Depending on the grade level of the participants, the two models can be tweaked to cover advanced topics in various approaches or be used for relatively simple observations. A more detailed explanation is provided following the explanation of the modules.

The effect of heat may activate thermal receptors in sensory neurons, resulting in them depolarizing and producing action potentials. Heat can also alter the resting membrane potential of cells, as illustrated in the concepts covered in **Chapter 1**, which considered the equilibrium potential of ions influenced by temperature.

Key Terms:

The following key terms will be used throughout the following modules. Participants should be encouraged to study the terms as needed. However, this list is not exhaustive, and participants should add additional terms to this list as needed.

- *Drosophila melanogaster*: a common animal model in the field of biology that is also referred to as "fruit fly." Fruit flies undergo several developmental stages over the course of their lives, starting as first instar larvae and ending as adult flies

- Electrical field: a region or space that contains electrically charged particles
- Thermal gradient: a region or space that contains zones of different temperatures
- Developmental stage: a period or stage in an animal's life. For instance, the second instar larval stage is a developmental stage of fruit flies
- Protein under/over-expression: the low or excessive amount of protein expressed
- Transient receptor potential ankyrin 1 (TRPA1): a Ca^{2+}-permeable thermosensory ion channel involved with temperature sensation in *D. melanogaster*
- (Sensory) receptor: a protein (e.g., ion channel) that senses external stimuli (e.g., temperature)

2.1. Module 1:
Exploring the Effect of Electrical Field on Behavior

Step 1:
Formulate a Hypothesis

Materials: paper/lab notebook, pencil, and/or computer
Instructions & Details: Before starting the experiments, the participants should write down their predictions for the organism tested in being attracted or repelled or not sensitive to an electric field. Also, participants should make predictions on the behaviors related to the voltage source. The participants should also make predictions in the conductive nature of the poles to be used as well as the liquid media. The participants can make predictions about how the electric field in the dish may occur depending on how the poles are placed and the media used.
Example:

Step 2:
Set Up the Electrical Field

Materials: Petri dish, filter paper, battery/DC power source, volt/multimeter, wires

Instructions & Details: The demonstrated electrical experiment exposes larval *Drosophila* to an environment in which there is a voltage difference between positive and negative poles at opposite locations in a conductive media. Electrical sources can be attached to metal material or a carbon rod, creating an electrical field within a Petri dish that the invertebrates interact with. Electrical sources can be any material with the ability to produce an electrical potential, such as a battery or a DC power source. Any battery type will suffice; however, the stronger the voltage, the stronger the electrical field.

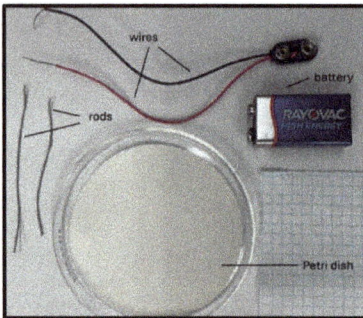

Figure 1*: Setup and voltage sources for generating an electric field. The relatively simple 5-volt battery with a clip to attach and two wires color coded (red + and black -) provides an easy set up. The use of soldering wire is convenient as it can be wrapped around the bare wire leads and it is very malleable to bending to fit in the dish and to remain in contact with the paper and media.*

If one wishes to use stronger voltage source, a DC power supply is an approach, but one needs to take care when handling the DC power supply, as it can generate a large voltage and current, which can be hazardous. Also, note that the small 5-volt battery can become very hot and potentially dangerous if the leads are left to touch each other, as this can result in the battery heating up to hot temperatures enough to burn skin or maybe result in a fire.

To take cautionary measures, make sure to disconnect the leads when not in use, and store without shorting the two leads to each other.

Various types of batteries can be placed into holders specific to their size. Some holders have wires exposed or tabs one needs to add some length of a wire leading to the conductive metal or carbon rod to be placed on the Petri dish. Each wire provides either a positive or negative charge. The insulation on the extension wire can provide places to anchor, the leads to a location in the dish. The poles in the dish need some length in the dish to provide a good electric field, not just a point source. We found that soldering wire works well as the wire is very pliable and lays flat on the surface in the dish; however, copper wire or carbon/graphite rods also work.

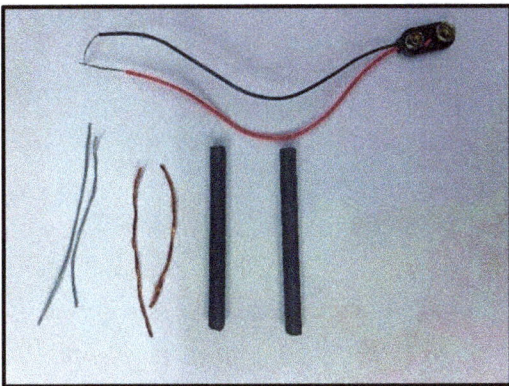

Figure 2: Example of battery leads and different types of conductive poles for use in the experiment.

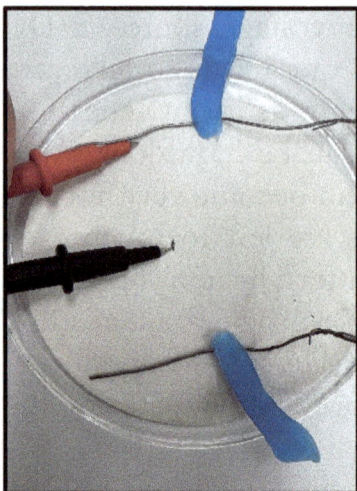

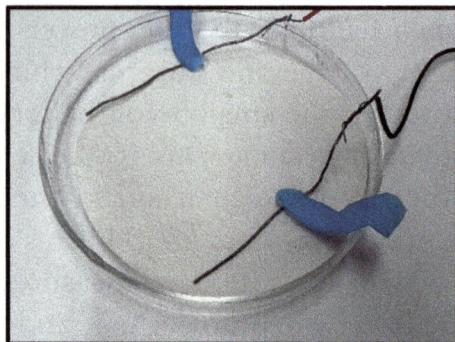

Figure 3: *The leads need to lay as flat as possible on the filter paper. Using wax or clay to anchor the leads work well. Only the soldering wire of about the same lengths are touching the paper in this figure.*

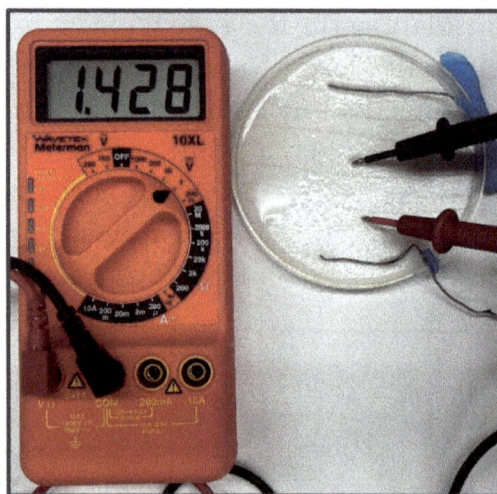

Figure 4: *Measures of electrical potential with a voltmeter. One can measure at set distances from the midline toward each pole and mark with a pencil or use a grid previously drawn on the filter paper the values obtained. It is easiest to keep the black lead as a reference in the center of the dish and move the red lead toward one pole or the*

other as the larvae could 1ˢᵗ be placed in the center of the dish to allow them to make a choice to which pole to avoid or be attracted to. The voltage is measure in a DC setting.

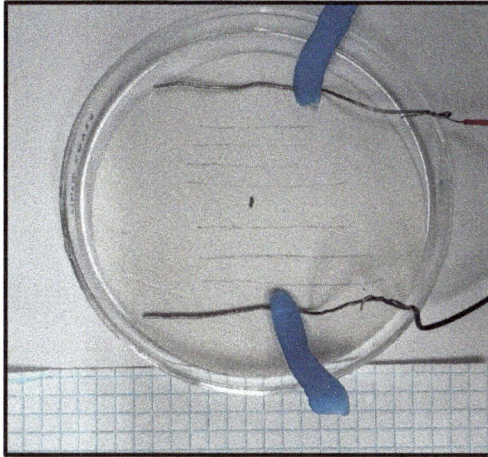

Figure 5: *A filter paper with grid lines drawn ahead of putting the conductive media in the dish. One could be more precise and even draw grid lines of set distances from a graph paper.*

Instructions & Details: A voltmeter can be used to quantify the electrical potential at various distances from the poles. Use one lead in the center between the two poles and move the other one at set distances towards one pole and repeat going the other direction to the other pole. If one uses a pencil, one can mark where the measurements are made to then correlate with the location of the larvae. One can also place one lead at one pole and move the other lead at set distances apart until reaching the other pole. Paper of some sort is used to provide a rough surface for the larvae to crawl on.

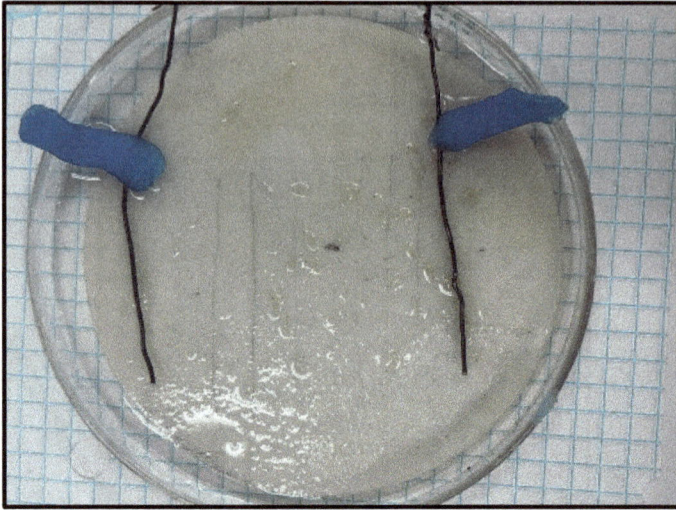

Figure 6: *If a thin filter paper is used, one might be able to see a graph paper under the dish or potentially use graphing paper that the ink will not run through. Cut the paper to fit the dish.*

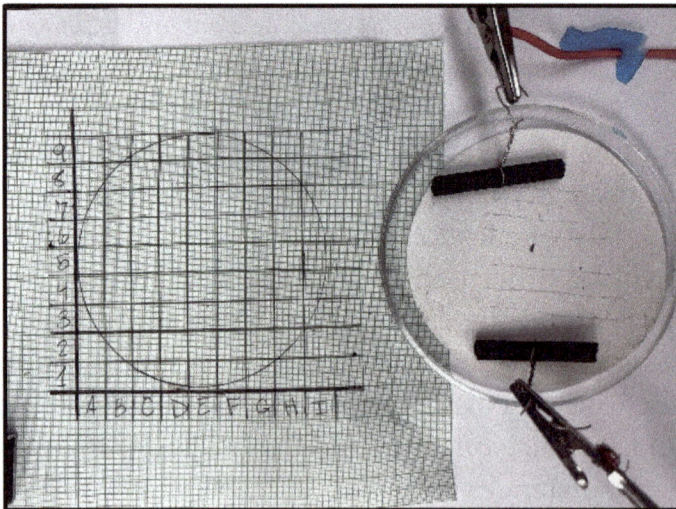

Figure 7: *One could use a grid paper marked with definitive markings for quantification such as letters and numbers for the two axes.*

Step 3:
Add Solution and Larvae
on the Filter Paper

Materials: electric field setup, *Drosophila* larvae, conductive solution

Instructions & Details: Once the wires are lined parallel in the Petri dish lined with filter paper, a conductive solution can be used to saturate the filter paper. This provides a medium to conduct an electric current. One can use sodas, sports drinks, any liquid with an ion (i.e. salt) content will work. The solution should just be enough to allow the larvae or other animals to walk or crawl on the paper without the body floating or being completely covered as to drown the organism. For example, *Drosophila* larvae breathe through the tracheal system and use the spiracles at the caudal end for gas exchange. These spiracles need to be able to reach out of the solution.

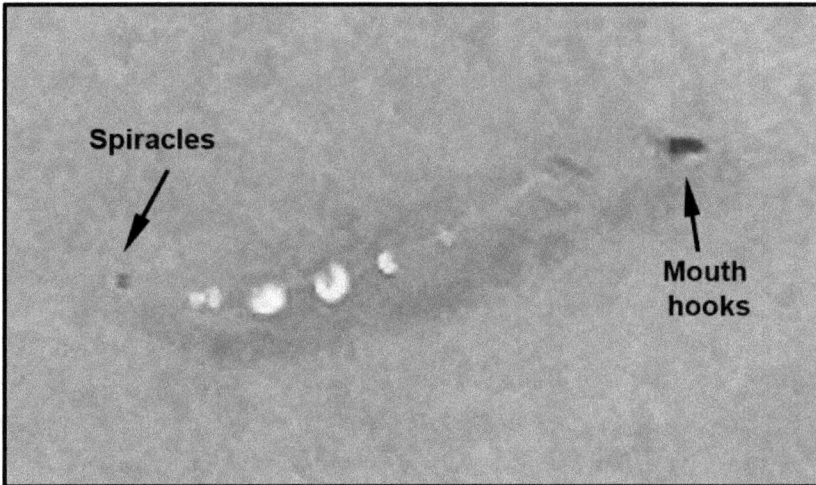

Figure 8: The spiracles are on the caudal end of the larvae. They need to be able to reach out of the liquid media. The mouth hooks move rapidly while eating.

Step 4:
Observe & Record Results

Instructions & Details: The larva behavior and movements are then observed for several minutes. Since the larval *Drosophila* are light sensitive, a paper towel can be used to cover the dish, blocking out light and allowing them to move to their preferences more quickly. It is possible to use dim room lighting.

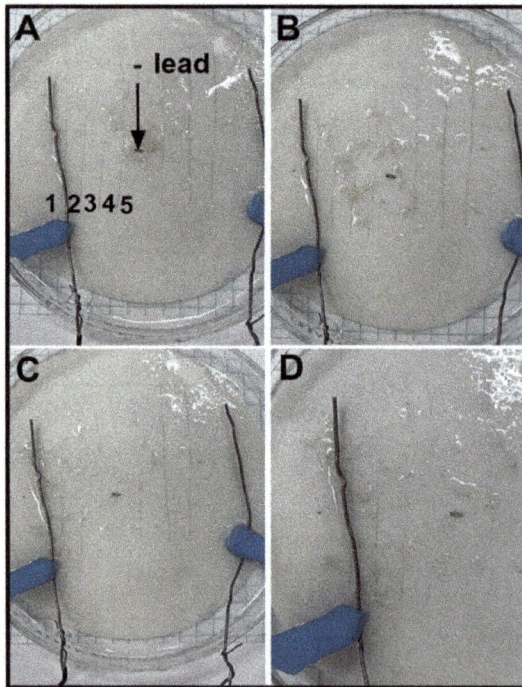

Figure 9: *Larval movement over time in an electric field. (A) The larvae are placed in the center of the dish while the 5-volt battery was connected. (B) After 10 minutes the larvae had already started toward the direction of the negative pole. (C) After 20 minutes many larvae had come in contact or close association with the negative lead. (D) An enlarge view of the lead shown in C. One can count the black mouth hooks for each strip or grid box for quantification.*

Instructions & Details (cont'd): To quantify the *Drosophila* larval behavior, one can use various approaches. One approach is to count how many larvae are on the wire or carbon rod over time. This allows for creating a data plot that shows the effects over time. Another approach is outlining set distances from the pole (1 cm) to the center of the dish, so each grid line can be used to count how many larvae are in each strip over time. This can be readily performed with a ruler or grid paper placed to the side of the dish. This can also be corelated with the voltage measures made.

Step 5:
Create Graphs/Tables for Data Analysis

Types of graphs as examples are illustrated below. Since each experimental set up may vary in the number of larvae used, a percent of total allows a normalization needed to compare between experiments performed. A table would also be appropriate if one does not have access to graphing supplies.

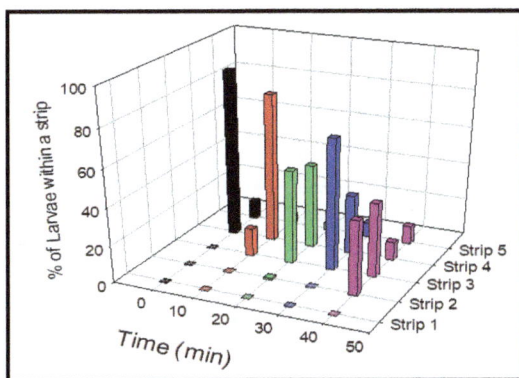

Figure 10: *Sample three-dimensional graph of the percent of the total larvae in each strip over time while exposed to the electric field.*

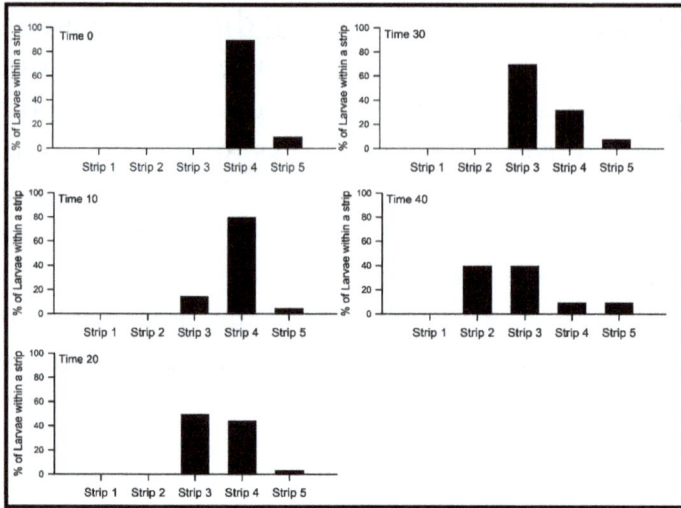

Figure 11: *Sample two-dimensional graphs of the percent of the total larvae in each strip over time while exposed to the electric field. Each 10 minute segment was graphed individually.*

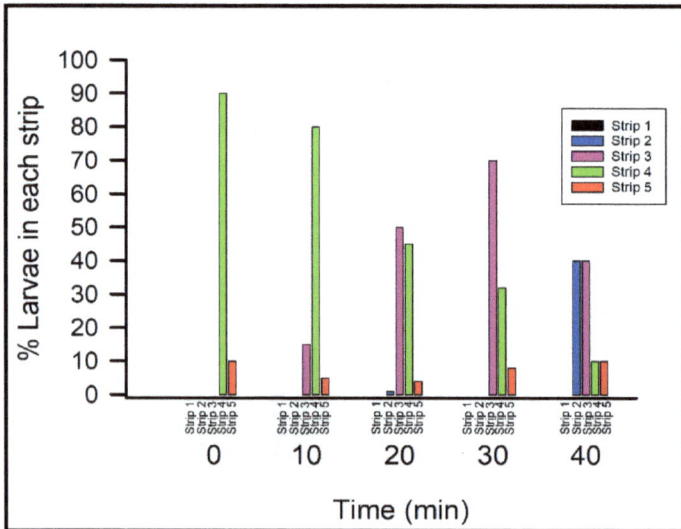

Figure 12: *Sample graph of the percent of the total larvae in each strip over time while exposed to the electric field. Each 10 minutes was graphed individually within each strip.*

2.2. Module 2:
Thermal Gradient on Behavior

Step 1:
Formulate a Hypothesis

Materials: paper/lab notebook, pencil, and/or computer
Instructions & Details: Before starting this module, participants should write down predictions for the organism tested in how they will respond to a thermal gradient. The participants should make predictions on the behaviors related to whether the surface is cold or warm. Participants could also make predictions about the response based on the developmental stage of the organisms if appropriate. Consider if one used larval *Drosophila* where a genetic manipulation could alter the expression of receptors that respond to heat in sensory neurons, how might the response vary in such larvae.

In this procedure an organism is exposed to a temperature gradient in which the organisms can choose their preferred location to the thermocline. In the examples provided, 3rd instar larval *Drosophila* are used. Larval *Drosophila* spend much of their time eating, so food cues will drive larvae to move. One approach to ensure they are content with the food source in various locations is to provide an environment with low level of dissolved yeast for larval *Drosophila* or a standard food source of corn meal but not too much to impede movement of the organisms.

Step 2:
Set Up the Thermal Plate

Materials needed: soldering iron, Petri dish, filter paper, copper (optional)

Instructions & Details: To apply a heat source, various approaches can be used. An inexpensive soldering iron is used with the ability to regulate the amount of heat. The heat source is placed targeting one specific area of a glass Petri dish lined with filter paper. (Note: the dish needs to be glass, as plastic will melt and can even catch fire). It is easier to provide a wider gradient by placing the heat source close to one side. The filter paper is moistened, but not with a thick layer, with a fluid such as water with dissolved yeast, soda, or apple juice. The temperature difference is amplified when the original dish is placed on top of another glass dish containing ice. If one has a soldering iron that cannot be controlled or the tip is too large, a copper wire can be wrapped around the end of the tip of soldering iron and the put one end of the wire on the filter paper as shown below.

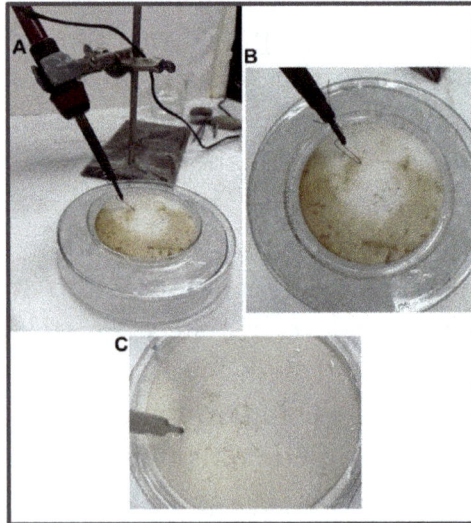

Figure 13: Creating a thermal gradient with a solder iron. (A-B) Some soldering irons have large tips and do not have a variable heating range. Thus, one might need to modify the probe by using a copper wire wrapped around the end if the soldering iron and use it as a

probe in the dish. (C) Some soldering irons have a variable setting3 for temperature control and a series of tips that can be used. A small tip works well placed directly on the filter paper and a low setting of 180 C degrees.

Step 3:
Record Plate Temperature

<u>Materials</u>: IR cooking thermometer (e.g., FLIR IR detector)

<u>Instructions & Details</u>: To record the temperature in various locations on the paper a thermometer that is standard with Vernier teaching kits (Surface Temperature Sensor along with LabQuest, can record 3 leads simultaneously) can be used. An IR cooking thermometer, or FLIR IR detector which connect to an I-Phone to determine the exact thermal level at any location in the dish maybe used.

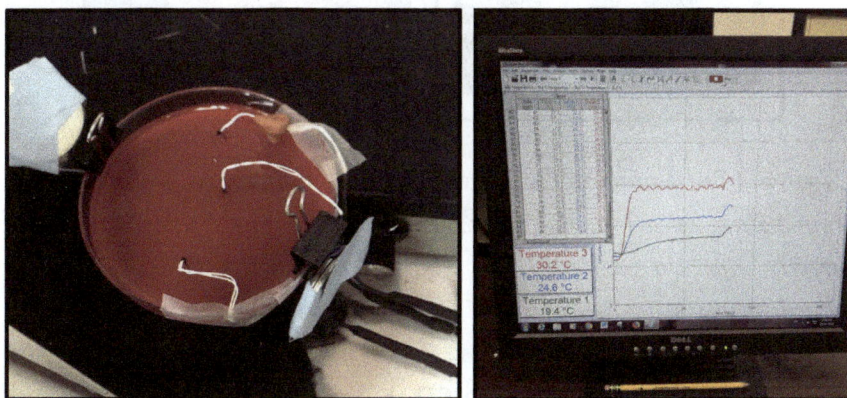

Figure 14: *Vernier student teaching kits come with software and thermal probes where 3 or more leads can be used simultaneously to measure a gradient across a surface. The software can obtain continuous measurements to examine variability over time or different experimental conditions.*

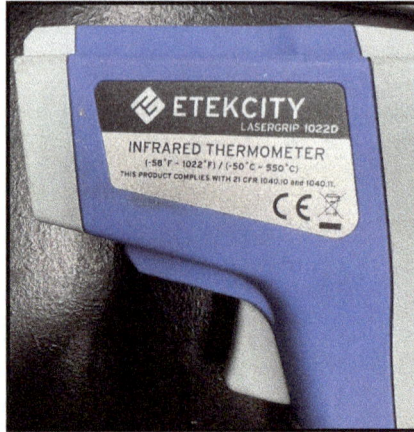

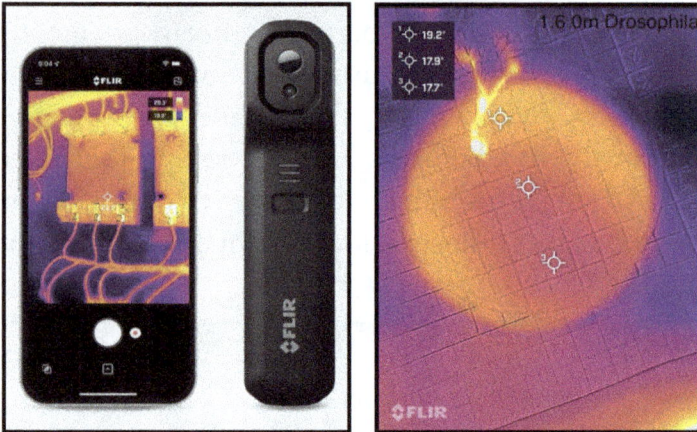

Figure 15: *Handheld devices such as an* IR *thermometer to measure temperature or a* FLIR *camera can be used. The* FLIR *can record directly to an* I-PHONE *with 3 measures being taken simultaneously.*

Step 4:
Add Drosophila to the Plate

Materials: *Drosophila* larvae

Instructions & Details: Add *Drosophila* larvae to the dish, in between the heat source point and cool outer edges. Since the larvae are light-sensitive, a paper towel or an aluminum sheet can be used to cover the dish, blocking out light and allowing them to move to their preferences more quickly. (Note: be careful the paper does not catch fire by the soldering gun). The cover can be removed and photos taken at set intervals of time (e.g., every 10 minutes).

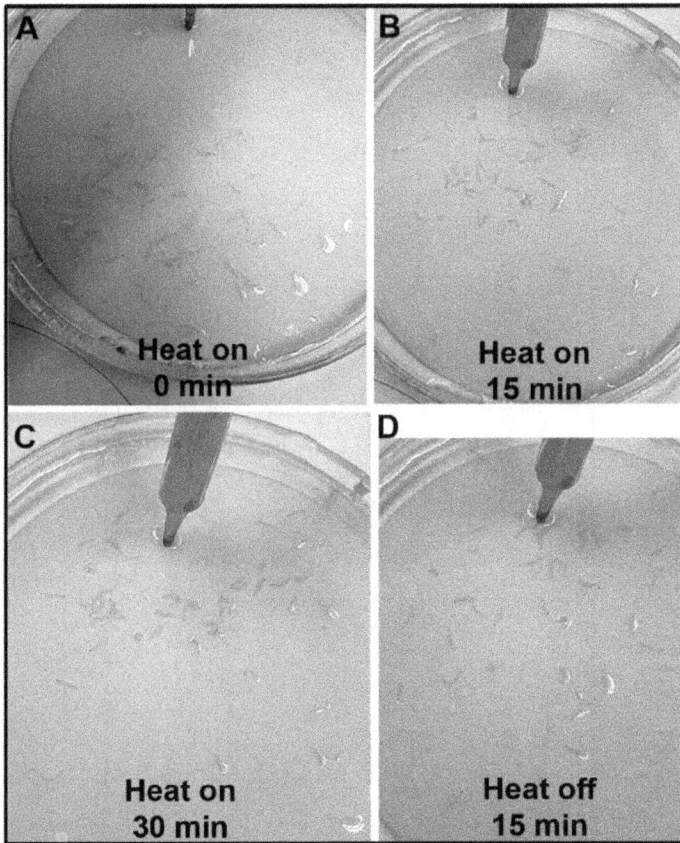

Figure 15: Larval movement over time in a thermal gradient. (A) Larve placed at the opposite end of the thermal source. (B) After 15 minutes the larvae were more closely aligned toward the heat source. (C) After another 15 minutes larvae were lined a boarder around the heat source. (D) When the heat source was turned off, after a few minutes the larvae approached and even touched the probe as well as scattered over a wider area.

Step 5:
Observe & Record Results

Instructions & Details: To quantify the effects, one can use various approaches. One approach is to count how many larvae are at varying distances from the heat source. A plot overtime could also be made. An approach in outlining set distances as rings from the pole (1 cm) with the center being the heat source could be used. This can be readily performed with a ruler or grid paper placed to the side of the dish with hand drawn outlines. This can also be corelated with the heat measures obtained.

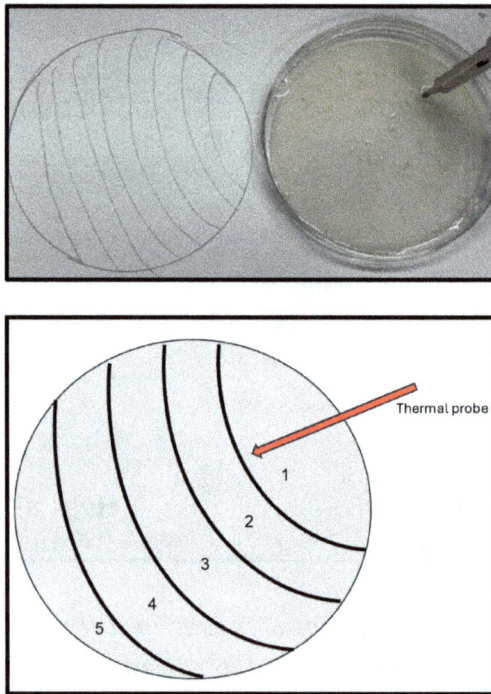

Figure 16: Hand drawn rings are an easy approach to quantify the behaviors over time. The drawing can be superimposed over the dish from photos or drawn prior to the experiments directly on the filter paper.

Types of graphs as examples are illustrated below. Since each experimental set up may vary in the number of larvae used a % of total allows a normalization to compare between experiments performed. Possibly a table would be appropriate if one does not have access to graphing supplies.

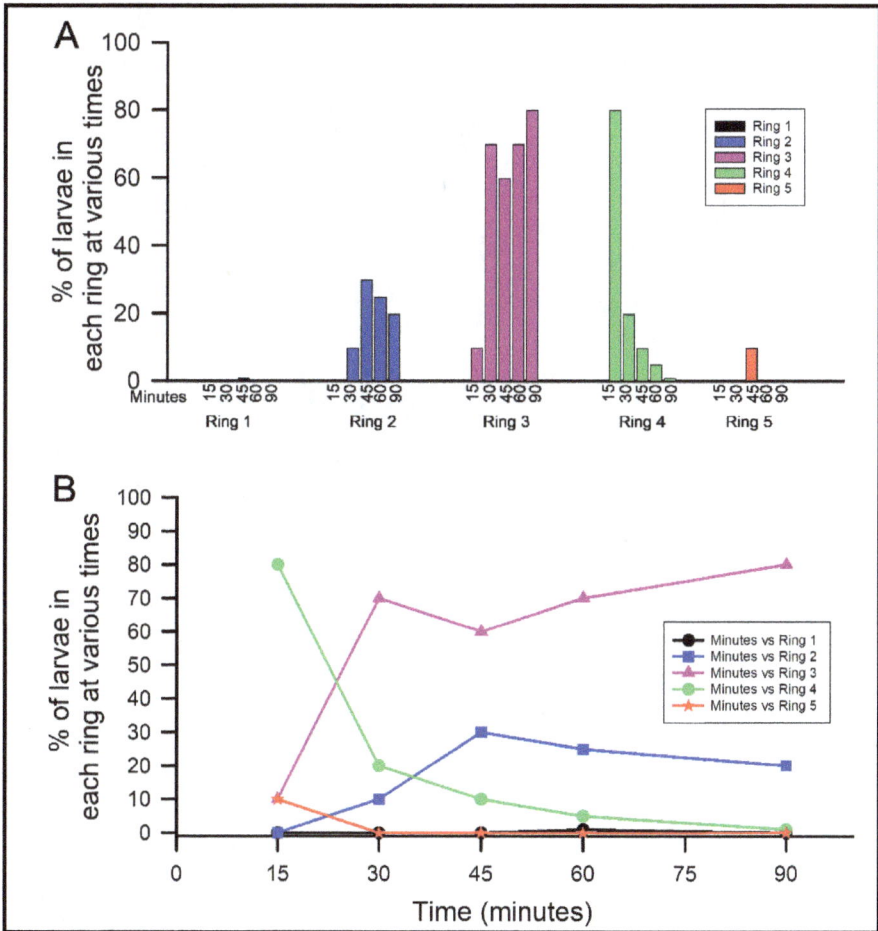

Figure 17: *Various approaches in quantifying the changes over time and preference of the larvae to particular thermal zones. (A) A % of larvae for each zone over time. (B) A % of larvae within each ring over time.*

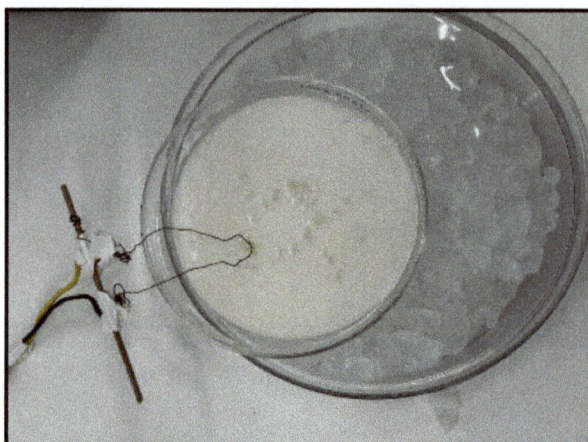

Figure 18: *Additional methods to create temperature gradient. A DC power supply with variable settings for voltage to heat a looped wire placed on the filter paper.*

2.2.1 Genetics: A Deeper Experiment on Thermal Gradient

If the participants have the ability to conduct genetic crosses with *Drosophila*, then one can over or under express the TRPA1 protein in subsets of sensory neurons. Transient receptor potential (Trp) family of channels are a broad category, with some acting as stretch activated channels as well as serving as thermal receptors. TrpA1 (TrpA - ankyrin) receptors are a known subtype that can expressed in sensory neurons to detect thermal sensation. TrpA1 receptor over expression was performed by crossing female virgins of UAS-TrpA1 from the from the Bloomington Drosophila Stock Center (BDSC stock # 26263) with males of ppk-GAL4 (class IV and class III dendritic arborization mechanosensory neurons) (BDSC stock # 32078).

If one wanted to reduce the expression of TRPA1 in cells the RNAi technique can be used to block any expression of the TRPA1 protein. Use of UAS-TrpA1-RNAi instead of UAS-TrpA1 can be used to cross lines. Tissue-specific RNA interference of TrpA1 can be performed using two different transgenic lines (BDSC stock # 66905) or (BDSC stock # 36780). The parental line of UAS-TrpA1 alone can be used for control comparisons. On the Bloomington Drosophila Stock Center homepage, navigate to the page for purchasing and enter the stock numbers to make a one-time purchase.

To generate larvae with both transgenes, collect three or four virgin adult female flies from one line and place them into a vial of food with two or three males from the other genotype. For more information about obtaining supplies for raising Drosophila: see Genesee Scientific web page (https://flystuff.com/).

A fun activity that can be conducted with an over expressing TRPA1 line is to use the juice of a smashed hot pepper such as the Carolina Reaper pepper which is 200x hotter than a Jalapeno pepper on one side of a dish lined with a filter paper and watch the larvae avoid that side of the dish. A dilution series can be applied in segments of the filter paper when the Petri dish is placed on top of an ice dish. They may prefer the region with more pepper. The reason the juice from the peppers will result in a response is that the ingredient of capsaicin and the related capsaicinoids will activate the TRPA1 receptor.

3. Learning Outcomes

3.1 Various Experimental Designs & Learning Outcomes for Module 1

- Participants could try other materials to examine electrical conductivity for the poles, such as aluminum strips from a soda can or even a steel wire. One can look up information on the conductivity of various metals. One can address how electricity travels in a solid material.

Learning outcomes can be focused on addressing electrical conduction and types of conductive media

- Different voltages can be tried with different types of batteries or a DC source and examine difference in timing or larvae to or away from the poles. This also allow different measures to be made on the voltage at various distances from the poles to be graphed depending on the voltage of the batteries.

Learning outcomes can address what makes a battery store charge and composition of batteries. In addition, addressing if organisms can sense voltage strength and research how they may sense the voltage. One can discuss cable properties to address why the voltage is dropping at various distances from the source. If careful mapping of the voltage sources are made, then a discussion can take place of why the voltages vary in various direction around the poles.

- Measures with same poles but different conductive media such as tap water, soda, apple juice to correlate the voltage measures on the dish with the media.

Learning outcomes how electricity travels in a liquid with salts and resistance of current. Again, addressing cable properties of current. Chemical makeup of the various media and what substance would be considered conductive (Glucose, Na^+Cl, flavors etc...).

- Participants could learn how to construct their own graphs with ruler and paper and think are various want to make measures as well as to report their results in the most effective way.

Learning outcomes are in determining how to collect data and to quantify the observations. The various approaches in graphing or presenting the data.

- Participants can try various types of organisms such as termites, Roly-polies, ants, meal worms, snails, etc., and discuss why differences in the responses occur depending on the organism.

Learning outcomes can be focused on how animals sense the electric field and differences among the organisms.

Multiple types of paper material (graph, filter paper, etc.) used to line the petri dish must be correctly sized, introducing mathematical concepts associated with the radius, circumference, and area of a circle.

Conductivity is dependent on the material used for the wires, creating lessons associated with electrical conductivity, malleability, and other character differences in metals.

Liquid mediums having the ability to conduct electric currents must be utilized, and this can be stock created or experimentally created, creating the opportunity for lessons on molarity, molality, concentration and its relationship to conductivity, polar/nonpolar molecules and their solubility and ionization.

3.2 Various Experimental Designs & Learning Outcomes for Module 2

Other heat sources include a DC power supply which an exposed tip of a wire can be used. The wire heats due to the high resistance with one end of the wire attached to a **+** pole and the other end on the **−** pole as shown (Figure 18).

Various levels of heat could be used and with and without the cold ice plate under the behavioral dish.

Multiple types of paper material (graph, filter paper, etc.) can be used to line the dish. Such measures can introduce mathematical concepts associated with the radius, circumference, and area of a circle. One could place the heating coil in the middle or to the side of the dish and draw a series of expanding rings of a giving distance apart for later measures of the larvae locations.

Learning outcomes can be focused on how heat is produced with electricity. A focus can be on how heat dissipates across the dish as well as how heat is detected by animals and drives then to seek a particular thermal zone. There are many topics that can be researched by the participants.

4.0 Comments From Past Student Participants

Students who have participated in the beta testing of this project have made suggestions for improving the research experiments. Some students suggested focusing on how far away the electric poles are able to be placed in the electrical model to still have an effect on the larvae. Other students considered altering the genes of the larvae that would make them more susceptible to the influence of thermal changes (TRPA1 receptors). Most students, however, dug into methods that would allow for the testing of both models simultaneously. One student suggested using a negative pole, which could be varied in voltage, on opposite sides of the thermal probe with the dish remaining on the ice platform. Student researchers would be able to analyze which stimulus (electrical or thermal) is favorable for attraction in the larvae. Another student commented on using the ice platform with the electric poles to determine if that might result in less sensitivity to the electric field, as neuronal activity may be dampened due to the cold environment. In addition, another student took genetically modifying the TRPA1 receptor a step further by performing the electrical experiment on these larvae and comparing results to the control to determine if a difference occurs.

5.0 Suggested Materials and Substitutions

I.   Petri dish.
     A.   Can be substituted using a small, shallow, clear container that allows for similar visibility to a petri dish.

II. Filter paper (sized to fit the petri dish).

A. Can be substituted with printer or graph paper cut into a circle to fit inside the petri dish. Purposely using this substitution could introduce math skills associated with the radius, circumference, and area of a circle. Using graph paper may prove as a superior medium due to the lines, ensuring the wires are placed in relatively equal distance later on.

III. Batteries.
A. The 5-volt battery as shown in the text can be substituted with other types of batteries (AAA, B, C, etc.), but it will impact the level of voltage. Using batteries with too high of voltage will have a risk factor, while using batteries with too low of a voltage will not have a significant impact in the experiment. Experimenting with different battery types could allow for lessons on voltage and conductivity associated with different batteries.

IV. Battery holder with wire.
A. Can be substituted with other types of battery holders with a wire depending on the battery used.

V. Wires
A. Can be substituted with other wire types (soldering, copper, steel, gold, etc.), but it will impact the level of conductivity. Purposely using different wire types could allow for lessons on electrical conductivity,

malleability, and other character differences in metals.

VI. Conductive liquid medium.
    A. Any liquid with the ability to conduct electricity can be utilized. Examples of conductive mediums are: electrolyte sports drinks, sodas, or sodium chloride aqueous solutions.
        1. Using a created aqueous solution such as sodium chloride could allow for lessons on molarity, concentration and its relationship to conductivity, polar/nonpolar molecules and their solubility behaviors, ionization, etc. This specific solution can be created by mixing sodium chloride (table salt) with water.
        2. Using a pre-made solution such as a sports electrolyte drink or soda would allow for lessons on molarity, but serve as a constant in the experiment-meaning if multiple experiments were run at once then the concentration would be perfectly equal since it was predetermined.

VII. Anchoring of wires to the dish.
    A. Wax or clay appears to work well. Also, clips can be substituted to hold the wires to the side of the dish if needed. The clips should keep from touching the conductive media.

VIII. Thermometer.
    A. Can be any thermometer of any price range. This can be substituted by simply feeling the wires, batteries, and container as the experiment progresses to check for heat. Time between cold to hot materials can be recorded to demonstrate heat is produced.

    Voltage meter.
    B. Can be any voltage meter or multimeter of varied price range. While this cannot be substituted, one can imagine the transfer of energy and "see" it when the invertebrates "line up" on their preferred level.

IX. Invertebrate insects.
    A. Can be any invertebrate insects. This experiment focuses on the behavior of *Drosophila melanogaster*, the common fruit fly, but other options of insects include: pillbug (*Armadillidium vulgare or Latreille*), termites (*Isoptera*), etc. could be used.

References

Bakshi, A., Patrick, L.E., Wischusen, E.W. 2016. A framework for implementing course-based undergraduate research experiences (CUREs) in freshman biology labs. The American Biology Teacher. 78(6):448–455.

Barbagallo, B., Garrity, P.A. 2015. Temperature sensation in Drosophila.
Current Opinion in Neurobiology. 34:8-13.
doi: 10.1016/j.conb.2015.01.002.

Coll, R. K., France, B., Taylor, I. 2005. The role of models/and analogies in science education: implications from research.
International Journal of Science Education. 27(2):183–198.
https://doi.org/10.1080/0950069042000276712

Crampton, W.G.R. 2019. Electroreception, electrogenesis and electric signal evolution.
Journal of Fish Biology. 95(1):92-134.
doi: 10.1111/jfb.13922.

England, S.J., Robert, D. 2024. Prey can detect predators via electroreception in air. The Proceedings of the National Academy of Sciences (PNAS), USA. 121(23):e2322674121.
doi: 10.1073/pnas.2322674121.

Esparza, D., Wagler, A.E., Olimpo, J.T. 2020. Characterization of instructor and student behaviors in cure and non-cure learning environments: impacts on student motivation, science identity, development, and perceptions of the laboratory experience. CBE Life Sciences Education 19:1-15.

Kang, K., Panzano, V.C., Chang, E.C., Ni, L., Dainis, A.M., Jenkins, A.M., Regna, K., Muskavitch, M.A., Garrity. P.A. 2011. Modulation of TRPA1 thermal sensitivity enables sensory discrimination in Drosophila.
Nature. 481(7379):76-80.
doi: 10.1038/nature10715.

Krajcik, J. S., Czerniak, C. 2014. Teaching Science in Elementary and Middle School: A Project-Based Approach (4th ed.). New York: Routledge.

Krajcik, J., Merritt, J. 2012. Engaging students in scientific practices: what does constructing and revising models look like in the science classroom?
Science and Children. 49(7):10–13.

Linn, M.C., Palmer, E., Baranger, A., Gerard, E., Stone, E. 2015. Undergraduate research experiences: Impacts and opportunities.
Science. 347(6222):1261757.
doi: 10.1126/science.1261757.

National Research Council. 2000. Inquiry in the National Science Education Standards: A Guide for Teaching and Learning. Washington, DC: National Academy Press.

NGSS Lead States. 2013. Next Generation Science Standards: For States, By States. Washington, DC: National Academies Press.

Norman, G.R., Schmidt, H.G. 1992. The psychological basis of problem-based learning: a review of the evidence. Academic Medicine. 67:557–565.
doi: 10.1097/00001888-199209000-00002.

Marx, R.W., Blumenfeld, P.C., Krajcik, J.K., Soloway, E. 1997. Enacting Project-Based Science.
The Elementary School Journal 97(4):341-358.

Tadres, D., Riedl, J., Eden, A., Bontempo, A.E., Lin, J., Reid, S.F., Roehrich, B., Williams, K., Sepunaru, L., Louis, M. 2025. Sensation of electric fields in the Drosophila melanogaster larva.
Current Biology 28:S0960-9822(25)00299-4.
doi: 10.1016/j.cub.2025.03.014.

Virtue, E.E., Hinnant-Crawford, B.N. 2019. "We're doing things that are meaningful": Student perspectives of project-based learning across the disciplines. Interdisciplinary Journal of Problem-Based Learning. 13(2):9.
doi: 10.7771/1541-5015.1809.

About the Authors

Youngwoo Kim

Youngwoo Kim is a high school junior who is currently attending the Gatton Academy of Mathematics & Science, a residential STEM program for gifted and talented high school juniors and seniors who have a passion in STEM fields. He works as a student researcher at the University of Kentucky biology lab under the guidance of Dr. Robin L. Cooper, studying membrane potential effects through computational simulations. He loves playing tennis and chess during his free time, and his dream is to build a fully function Iron Man suit one day.

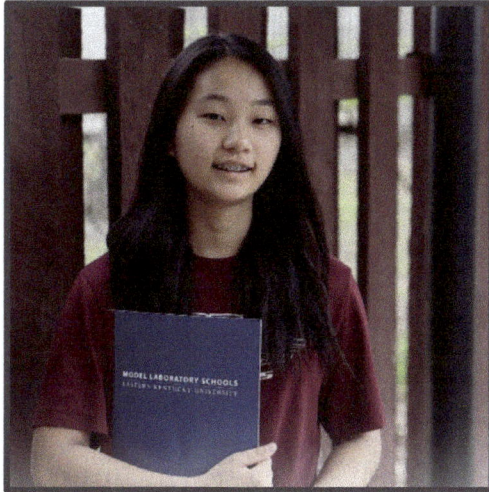

Jiwoo Kim

Jiwoo Kim is a class of 2027 high school student who is currently a sophomore at Model Laboratory School and a rising junior at The Gatton Academy of Mathematics & Science. She works as a student researcher in Dr. Robin Cooper's neurobiology lab in the T. H. Morgan Building at the University of Kentucky. Her past research projects have focused on the sensory and motor systems of *Drosophila* larvae and crayfish, as well as other topics like membrane potential. In her free time, she enjoys reading, writing, and playing sports.

Elizabeth Womack

Elizabeth Womack is a freshman attending the University of Kentucky majoring in honors pre-medical neuroscience, minoring in psychology, and obtaining a certificate in medical behavioral sciences. In addition, she is a part of the Lewis Honors College, Singletary Scholars 2028 Cohort, HealthCare Cats, and Vice President of Recruitment in the Kentucky Rural Health Association. She works as a student researcher in Dr. Robin Cooper's neurobiology lab in the T. H. Morgan Building at the university. She is from Flemingsburg, Kentucky and enjoys traveling, new experiences, baking, and spending time with loved ones.

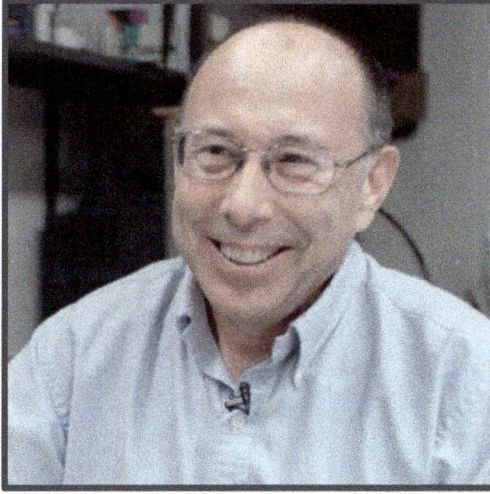

Robin L. Cooper

Dr. Robin L. Cooper obtained a dual B.S. in Chemistry and Zoology from Texas Tech University in 1983 and a Ph.D. in 1989 in Physiology from the School of Medicine, and postdoctoral training (1989-1992) at the University of Basel, School of Medicine, Basel, Switzerland. His second postdoctoral stint (1992-1996) was at the Department of Physiology, University of Toronto, School of Medicine, Toronto, Canada. In 1996, he joined the Department of Biology at the University of Kentucky and is now a Professor. He also obtained a BSN in nursing in 2012 and practiced as an RN from 2011 to 2017. He has received several teaching awards and continues to mentor students in research based activities and publishing for peer review. In his spare time, he FaceTimes with his first grandchild, Rose, and cycles the backroads of Kentucky.

Dr. Josh Titlow

Dr. Josh Titlow earned a B.S. in Chemistry from Marshall University. His Ph.D in Biology at University of Kentucky was advised by Professor Robin Cooper. After a postdoctoral fellowship in the University of Oxford Biochemistry Department, Dr. Titlow became a Technical Advisor and Program Coordinator for Delta Tissue, a $55M Wellcome Leap Program that developed pivotal spatial proteomics technologies and human tissue profiling datasets in Triple Negative Breast Cancer, Glioblastoma, and Tuberculosis. He is currently the Sr. Director of Substance Use Disorder and Crisis Divisions at Westbrook Health Services, and has published over 20 articles in neuroscience, advanced microscopy, and RNA biology. Born and raised in the Appalachian Mountains, Josh is also active in the outdoors and in bluegrass music communities.

To Contact the Authors:

Youngwoo Kim
ywkim17@outlook.com

Jiwoo Kim
kjiwoo468@gmail.com

Elizabeth Womack
eawo230@uky.edu

Robin Cooper
rlcoop1@UKY.edu

Josh Titlow
jstitlow@gmail.com

To Order Book Copies:

www.Lulu.com

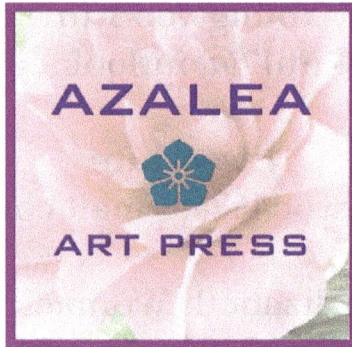

To Contact the Publisher:

Karen Mireau
Azalea.Art.Press@gmail.com

www.ingramcontent.com/pod-product-compliance
Lightning Source LLC
Chambersburg PA
CBHW052117090426
42741CB00009B/1852